INFORMAL LOGIC
The First International Symposium

INFORMAL LOGIC

THE FIRST INTERNATIONAL SYMPOSIUM

Edited by

J. Anthony Blair
and
Ralph H. Johnson

University of Windsor
Windsor, Ontario

Edgepress   Inverness   California

Library of Congress Cataloging in Progress
Card Number 80-67674

International Standard Book Number ISBN 0-918528-09-7

© 1980 by Edgepress, Box 69, Pt. Reyes, CA 94956, U.S.A.

Manufactured in the United States of America

TABLE OF CONTENTS

Preface   *vii*

Introduction   *ix*

THE EMERGENCE OF INFORMAL LOGIC

1. **The Recent Development of Informal Logic**   3
   Ralph H. Johnson and J. Anthony Blair

THE INFORMAL FALLACIES

2. **The Nature and Classification of Fallacies**   31
   Howard Kahane
3. **Petitio Principii and Argument Analysis**   41
   Douglas N. Walton

FORMALISM: PRO AND CON

4. **What Is Informal Logic?**   57
   John Woods
5. **Arguments That Aren't Arguments**   69
   Peter A. Minkus

PEDAGOGY AND PRACTICE

6. **Can the Ability to Reason Well be Taught?**   79
   Robert Binkley
7. **Advertising: Its Logic, Ethics and Economics**   93
   Alex C. Michalos
8. **Evaluation of Informal Logic Competence**   113
   Thomas N. Tomko and Robert H. Ennis

PERORATION

9. **The Philosophical and Pragmatic Significance of Informal Logic**   147
   Michael Scriven

Afterword   161

A Bibliography of Recent Work in Informal Logic   163
Ralph H. Johnson and J. Anthony Blair

To the memory of
Rev. Eugene R. Malley, C.S.B.
1924−1980

PREFACE

The papers collected in this volume, with one exception, were presented at the first international Symposium on Informal Logic held at the University of Windsor on June 26–28, 1978.

The basic premise behind the calling of the Symposium was a simple one: the time was ripe. Interest in informal logic was growing, and rapidly. Courses in informal logic or critical reasoning were springing up at an astonishing rate across North America, and work on informal logic in the journals was increasing markedly. At the same time, there was little if any contact between philosophers working and teaching in the field. In fact, what was remarkable about the proliferation of informal logic courses and writing was that it appeared to exhibit a sort of spontaneous, unconnected eneration. Another feature of these developments was a paucity of broadly-focused theory. (The theoretical work in the journals was largely directed in a scattered way at various informal fallacies.) Hence it appeared that there would be great value in bringing together some of the leading thinkers in informal logic for an exchange of ideas. The hope was that the Symposium would serve to highlight the present status of informal logic and provide nurture for its further development.

The editors were impressed by the noticeable excitement among those in attendance that the Symposium generated. Here for the first time, it seemed, was an opportunity to share ideas about an enterprise that the participants had hitherto been working on with a sense of being alone in the wilderness. Important issues came into clearer focus as a result of the papers and the discussions afterwards. It was generally agreed that it was important to publish the proceedings, so that others working in the field who did not attend would be able to share in this experience. A result of the discussions among the participants was the decision to ask Thomas Tomko and Robert Ennis to contribute a paper to the proceedings on the evaluation of efforts to teach informal logic.

This, then, is the background to the publication of these proceedings. It is our hope that this volume will further the development of informal logic as an independent and important field of inquiry.

Thanks are due to the Canada Council (now the Social Sciences and Humanities Research Council of Canada), the University of Windsor, and Mr. Richard C. Webster for funding the Symposium. We would especially like to acknowledge the support and encouragement of Dr. J. F. Leddy, then President, and the late Professor Eugene Malley, then Dean of Arts, of the University of Windsor. We are grateful to the Research Board of the University of Windsor for a grant in aid of publication, and to Dean Paul Cassano of the Faculty of Arts for his support. Special thanks are due to Violet Smith for her patience and perfectionism in typing the manuscript, and to June Blair for reading the proofs and other editorial assistance. We appreciate also the cooperation of the staff at Edgepress, particularly Sienna S'Zell.

It may appear unseemly to single out for special thanks one of the contributors to this book, especially when all have shown supererogatory patience and cooperation in support of the project. However, we wish to acknowledge a particular debt of gratitude for the enthusiasm and encouragement of Michael Scriven, without whose backing this volume might well not have appeared, and certainly would not have appeared as promptly and attractively.

JAB
RHJ

INTRODUCTION

I

The label "informal logic" means different things to different people. To many it refers to the lists of informal fallacies and the various descriptions and classifications of these fallacies—the tradition which began with Aristotle's *On Sophistical Refutations* and which has most recently been examined critically by C. L. Hamblin in his monograph, *Fallacies* (1970). To others it designates the subject matter of a certain sort of introductory logic course (or a segment of such a course) which employs various non-formal techniques (often but not always including the study of fallacies) to try to teach elementary reasoning skills. To still others, especially recently, it has come to mark off a field of logical investigation distinct from formal deductive logic. No doubt there are other ways in which "informal logic" is used. Indeed, we expect some would consider the label a contradiction in terms, for since they understand by "logic" the study of formal systems, informal logic would be a logical impossibility.

In the face of such disparate conceptions of informal logic, we would like to describe the interests and the focuses that motivated the papers which make up this volume. In the process, we outline our conception of the inquiry this label designates.

How is this field to be defined? There are at least two ways in which an area of inquiry might be characterized: in terms of the approach or methodology employed in it, and in terms of its subject matter. We think informal logic is best specified in terms of its subject matter, for as the papers in this volume show, there is no single approach shared by everyone whose work may be identified as belonging to it. At the same time, we must warn that there is no uncontroversial way to demarcate precisely the subject matter of informal logic. The reason for this is clear enough. The field is simply too undeveloped at this stage for a clear definition to be possible. The kinds of questions being raised, the kinds of problems being addressed, represent a diverse range of issues. Nevertheless, when they are set down side by side there emerges a coherence—admittedly loose—that can be seen to constitute a broad but distinctive area of inquiry.

We submit the following list of attitudes as characterizing the informal logic point of view. These are drawn from the general literature and, as the reader will find, they also are well represented in the papers collected in this volume.

1. A focus on the actual natural language arguments used in public discourse, clothed in their native ambiguity, vagueness and incompleteness.

2. A commitment to the study of argumentation as a dialectical process.

3. Serious doubt about whether deductive logic and the standard inductive logic approaches are sufficient to model all, or even the major, forms of legitimate argument.

4. A dissatisfaction with formal logic as the vehicle for teaching skill in argument evaluation and argument formation.

5. A conviction that there are standards, norms, or advice for argument evaluation that are at once logical—not purely rhetorical or domain-specific—and at the same time not captured by the categories of deductive validity, soundness and inductive strength.

6. A desire to provide a complete theory of reasoning that goes beyond formal deductive and inductive logic.

7. An interest in expanding the study of reasoning to include, besides argumentation even broadly conceived, such activities as problem-solving and decision-making.

8. A conviction that the informal fallacies constitute a legitimate basis for logical investigation. This conviction is often accompanied by an acknowledgment of the lack of any coherent theoretical account of the fallacies, and a commitment to provide such an account. There is, further, a growing belief that the theory of argument will be inadequate until it provides a framework for interpreting the informal fallacies.

9. A focus on the actual component skills of critical thinking, and hence an attempt to formulate a clear and operational concept of critical thinking or reasoning.

10. An orientation that treats the teaching of reasoning skills as a key part of education, integral to comprehensive language use skills and to preparation of youth for responsible social and political roles.

11. A belief that theoretical clarification of reasoning and logical criticism in non-formal terms has direct implications for such other branches of philosophy as epistemology, ethics and the philosophy of language.

12. An interest in all types of discursive persuasion, coupled with an interest in mapping the lines between the different types and the overlapping that occurs among them.

While not all informal logicians would pledge allegiance to all items on this list, what does seem to be shared is an interest in the theory and pedagogy connected with reasoning and argumentation in directions outside the scope of, though not necessarily incompatible with, the work in formal logic which has been the paradigm of "logic" over the past century.

II

We have no doubts whatever that informal logic is emerging as a separate field of philosophical interest. Witness the deluge of new textbooks that have appeared since 1970 and which, at the end of the decade, shows no signs of abating. Witness the increasing numbers of journal articles in the field. The bibliography at the end of this book supports these claims. In short, there is evidence of a developing philosophical attention to informal logic.

For present purposes, the proof of the pudding is in the papers presented to the first international Symposium on Informal Logic which, together with an additional invited paper, are collected here in revised form. These papers provide a cross-section of work in informal logic. In the process, they serve as further elaboration of the ostensive definition of the field which we offered above. We shall take up the remainder of this Introduction with brief summaries of the papers, together with suggestions about how they relate to one another, including notes about disagreements that exist among them.

We have organized the papers as follows. Chapter 1 introduces the volume by tracing the emergence of informal logic as a field of inquiry and pedagogical attention over the past twenty-five years, and particularly the past decade. It reviews and organizes the growing informal logic literature. Chapters 2 and 3 deal with the topic of informal fallacies, which can be viewed as perhaps the most important historical source of current work in informal logic. Chapters 4 and 5 represent a debate between outlooks within informal logic which is in many ways central to its conception—the issue is the place of formal methods in the field. Chapters 6, 7 and 8 address in different ways the interest in pedagogy and practice which has been the catalyst for much of the contemporary attention given to informal logic. Chapter 9, the keystone of the collection, depicts the many implications of this new field. In the Afterword the editors mention some of the developments that have occurred since the Windsor Symposium was held. Appended to the volume is a Bibliography of monographs, articles and textbooks in informal logic which have appeared over the past twenty-five years.

The Emergence of Informal Logic. Chapter 1. In their paper, "The Recent Development of Informal Logic," Ralph Johnson and Anthony Blair trace the remarkable growth of literature in informal logic, especially over the past decade. They outline the three important monographs by Toulmin, Perelman and Olbrechts-Tyteca, and Hamblin, that they think require attention by anyone who wants to do theoretical work in the field. They chart the journal literature, grouping articles by topic and sketching their contents. Finally, they offer an extended discussion of the textbooks in informal logic. They contend that these texts can be divided into two generations. They argue that in first-generation texts (pre–1970) what they call the "global paradigm" typified by Copi's *Introduction to Logic* is predominant. These texts are organized into three parts, dealing with (1) language, (2) deductive logic, (3) inductive logic. In the decade since 1970, they argue, a second "critical thinking" paradigm emerges, and they sketch its salient features, documenting their case with reference to the more recent informal logic textbooks. They append to their paper a list of the problems and issues in informal logic that they think need to be addressed.

The Informal Fallacies. Chapter 2. In his paper, "The Nature and Classification of Fallacy," Howard Kahane presents the notion of fallacy as part of a dialectical pair of concepts—cogent vs. fallacious argument (reasoning)—and argues that the idea of a fallacious argument cannot be reduced to the notions of validity or soundness. Cogent reasoning, he contends, must meet three criteria; it must be valid, move from warranted starting points, and use all the relevant information at our disposal. Thus a fallacious argument is one which fails to satisfy one of these requirements. Such failures can be classified in two different ways: either on logical grounds or psychological grounds and Kahane discusses each of these modes of classification in some detail. Because the concept of fallacy is central to informal logic, and because Kahane has here provided some interesting ideas about the nature of fallacy and how fallacies may be classified, this paper is an excellent starting point for those wishing to sample the sorts of theoretical and conceptual issues confronting informal logic.

Chapter 3. Of all the informal fallacies, perhaps none has so vexed philosophers and logicians as has the *petitio principii*. How is it to be understood as a deficiency of argument? This is the question Douglas Walton addresses in his paper, "*Petitio Principii* and Argument Analysis." Walton's paper is intended as part of his and John Woods's general program to develop theoretically adequate models of the informal fallacies, and his larger thesis is that "some general understanding of the underlying concepts of argument that are involved in the major informal fallacies" is imperative. In tracing the history of the "Standard Treatment" of *petitio*, he finds "a pair of dualisms." The first is a controversy over whether the fallacy is an epistemological or a dialectical phenomenon. The second is a dispute over whether an "equivalence" or a "dependency" relation exists between premises and conclusion in circular arguments. Walton's critical review of the various approaches to *petitio* is not only interesting in itself, but has implications for the theory of argument as well. Walton sketches the elements of such a theory and shows how it yields an analysis of the *petitio* explained in terms of digraph modelling borrowed from graph theory.

Formalism: Pro and Con. The topic of Chapters 4 and 5 is formalism. To be more specific, one of the crucial questions which confronts informal logic is its relationship with formal logic. Historically, formal logic has been far more adequately funded by theory than has informal logic. The question is this: To what degree should informal logic strive to make use of the methods and procedures developed by formal logicians? The two papers in this section take diametrically opposed views on this question.

Introduction

Chapter 4. In his paper, "What is Informal Logic?" John Woods first outlines the essential features of formal systems and mentions the benefits of such systems, namely "richer appreciation of [their] subject matter" and an understanding of their "own expressive and demonstrative limitations and capacities." He argues that it follows that "being a mathematical system is not necessarily a liability for a theory of the fallacies," while agreeing that being such a system is not necessary for the fallacy theory to be "instructive or even deeply correct." Woods grants that the construction of *axiomatic logistic systems* may not be possible, or illuminating, for notions needed for fallacy theory. However, he insists, the quite distinct claim that fallacy theory cannot benefit from *formal treatment* does not follow: a distinction between formalization and formal methods is crucial.

Woods argues, citing his own work with Walton as evidence, that the use of formal methods can illuminate the informal fallacies. Its limitations are due to the undeveloped state of the formal models, on the one hand, and of the theoretical understanding of the fallacies, on the other. Woods goes on to defend the stronger theses that the mature theory of the fallacies will in fact be "a branch of formal theory that is essentially extralogical in major respects" and hence that informal logic, at least as understood as the theory of the informal fallacies, is *not* logic, although it includes "quite a bit of logic," and moreover that unless it is (as he believes it to be) amenable to formal treatment, it is not even fully deserving the name of theory.

In an Appendix to his paper specially written for this volume, Woods illustrates his contention that formal theories are useful in the analysis of informal fallacies by working through attempts to give formal treatment to the concept of the part-whole relation which is central to an analysis of the fallacy of **Composition and Division**. (One can also read Walton's paper on *petitio principii* as a similar object lesson.)

Chapter 5. Peter Minkus's paper, "Arguments That Aren't Arguments," represents a diametrically opposed approach to the informal fallacies and informal logic in general to that taken by Woods (and shared by Walton). Arguing from an avowedly Wittgensteinian point of view, Minkus claims that the sort of hankering for rules and generality which accompanies the spirit behind the use of formal methods cannot but in the end serve to defeat the aims of clarity and the avoidance of confusion. In doing logic, both formal and informal, Minkus warns, we tend to become fond of rules without qualifications, and we forget that these rules have only the force of the cases from which they were generalized behind them. Minkus dubs formal models "panarchic," and sees underlying them an infatuation with universality and a forgetting of language-variety. The unchecked yearning for formal principles he views as a form of pedantry.

What is exhibited in the confrontation between Woods and Minkus is the clash between two fundamentally different conceptions of the nature of philosophical clarification. Informal logic, it is clear, cannot avoid becoming another battlefield on which will be fought out a broader philosophical contest.

Pedagogy and Practice. The three papers in this section are disparate in focus; however once their contents have been outlined it should be clear how each in its own way relates to the teaching of informal logic.

Chapter 6. Robert Binkley argues in his paper, "Can the Ability to Reason Well be Taught?" that indeed it can, but to understand why requires an understanding of what is being taught and how it might be taught. He distinguishes between the question whether reasoning can be taught at all, and the question whether it can be best taught by informal logic courses. He sees this as in part an empirical question, and bemoans the lack of needed data. (In Chapter 8 Tomko and Ennis address directly the problems of obtaining such data.) Binkley identifies the ability to reason well as having two components: first, a skill, and second, a desire to reason well and a respect for good reasoning—"love of reason" he calls it. This combination constitutes "rationality," which is clearly a virtue. Thus Binkley is brought to the analogy between his question and Socrates's question in the *Meno*, "Can virtue be taught?" around which he has structured his paper. Binkley works through a series of parallels between his quest and that of Socrates, and from this emerges his conclusion that "the critical reasoning course has three aspects, the training of logical intuition by drilling, the pursuit of theoretical logical insight by some feasible substitute for Socratic dialectic, and the whole to be conducted in such a manner what it will conduce to the love of reason."

Binkley closes his paper with a series of practical suggestions about how such a critical reasoning course might be done.

Chapter 7. Many reasoning courses aim to teach the student how to critically analyze a variety of forms of discursive persuasion, from philosophical argument, through political rhetoric, to advertising. The case of advertising is an interesting one, since at times the model seems that of argument and at times that of pure nonrational conditions. Moreover, this is a domain which strikingly illustrates the impossibility of practicing rational criticism in abstraction from knowledge of subject matter and context.

Alex Michalos's paper, "Advertising: Its Logic, Ethics and Economics," is an attempt to grapple with this complex phenomenon. He begins by arguing that it is rational for advertising to be used to persuade the public, given the special sense in which advertising is a "public good." At the same time, given the insignificant differences between many products, advertising is often forced to take on "the logically and morally outrageous task" of trying "to persuade people to differentiate indistinguisables and to prefer one to another!" With the stage thus set for the rational critic to expect deceptive advertising, Michalos proceeds with three discussions: first, a case study of deceptive advertising by government, second, a discussion of the phenomenon of subliminal advertising (where the key issues are empirical, not logical, he argues), and third, an extended dialectical interchange confronting an attempt to defend contemporary advertising against some of the criticisms levelled against it. We would draw the reader's attention to this last section of Michalos's paper. It represents an excellent model for the sort of rational criticism that Binkley argues it should be the objective of reasoning courses to teach. Furthermore, it demonstrates graphically that reasoning skills cannot function in a vaccum, but are skills-in-reasoning-about ethical, economic, political and other issues.

Chapter 8. How do we tell when our teaching of informal logic skills has been successful? How do we measure how much and how well a student has learned? Facing these questions squarely forces us to answer others. What precisely is it that we are, or that we should be, trying to teach? How should we be trying to teach it? The absence of any sort of clear answers was remarked on at the Symposium. As a result, Thomas Tomko and Robert Ennis, who attended the Symposium and who are members of the Illinois Rational Thinking Project (at the University of Illinois, Urbana-Champaign), were asked to write a paper for this volume reviewing the testing and other evaluation materials that exist. Their paper, "Evaluation of Informal Logic Competence," is their response to that request.

Tomko and Ennis's paper is a meticulous and critical review of available critical thinking tests, of test theory, and more. They distinguish two standard kinds of testing—testing achievement relative to others taking the test, and testing achievement relative to fixed standards. They spell out the norms used for assessing tests: how consistent the test is, and how well it measures what is being tested; and they break these down into the variety of components employed. At the same time they point out problems, practical and theoretical, inhabiting different parts of test theory. Using their critical discussion of test theory as background, Tomko and Ennis then offer advice about how to check out a test offered on the market, and how to construct a reliable test of your own. They underline the limitations, dangers and traps to be aware of and avoided. The paper includes a discussion of other kinds of evaluation besides testing, and calls for long-term follow-up evaluation. Throughout the paper the focus is on informal logic in three distinct respects. Tomko and Ennis discuss existing tests, test theory, test selection and test construction from the point of view of the use of tests for evaluating informal logic skills. Concurrently, they identify critical points at which informal logicians can contribute to the clarification and refinement of test theory and application. Finally, they keep before the reader questions about the nature of informal logic and the objectives of critical reasoning courses that require further theoretical and pedagogical attention. The paper ends with a useful summary of the ground it covers. There is continual reference to the testing literature throughout, and the appended list of these sources is an excellent bibliography of such literature.

Peroration. Chapter 9. Michael Scriven's paper, "The Philosophical and Pragmatic Significance of Informal Logic," is a frankly partisan and inspirational rallying cry. It is Scriven's conviction that the informal logic "Movement" represents a dynamic new departure with immense significance for philosophy in general as well as for pedagogy. In the first part of his paper he lists its philosophical implications. It portends an end to the reign of formal logic and the recall of logic to its proper task, the study of argument.

This in turn promises to liberate other areas of philosophy which have been encumbered by a misplaced allegiance to formal logic. The study of the logical foundations of informal logic will produce concepts that will lead to new and fruitful ways of solving old problems (e.g., in the philosophy of science, in ethical theory, and even in formal logic itself). Scriven sets out a number of practical implications of informal logic in the second part of his paper. He sees it having a crucial role to play in developing adequate achievement tests. (This point is amplified by Tomko and Ennis.) Informal logic has widespread and far-reaching application to pedagogy. It should become a factor in developing the "basics" curriculum; it should be part of teacher training; it should inform the teaching of English composition; it should be brought to bear in teaching reasoning in science. Scriven argues that informal logicians need to get involved in the teaching process beyond just writing textbooks. He sees for informal logic the role of preparing responsible citizens, politicians, journalists, lawyers, for it is socially and politically important for these people to reason well, and reasoning must be taught. He holds that progress with the conceptual foundations of informal logic will connect directly with educational research. Finally, Scriven urges informal logicians to actively promote the field—by organizing meetings, promoting the *Informal Logic Newsletter*, engaging in the intellectual development of the field by writing journal aritcles, creating a journal, publishing resource anthologies. Scriven, in short, sees the emerging informal logic activities as an important movement with wide ramifications.

THE EMERGENCE OF INFORMAL LOGIC

THE RECENT DEVELOPMENT OF INFORMAL LOGIC

Ralph H. Johnson
J. Anthony Blair
University of Windsor

The purpose of this chapter is to provide an overview of the recent development of informal logic. Part I is an introduction consisting of two brief sections: (A) a short historical background; (B) a statement of our approach. Part II is a survey of the developments in informal logic over the past 25 years. In Part III, after summarizing these developments, we attempt to formulate the central issues and problems with which informal logic must deal.

PART I: INTRODUCTION

(A) Background. Logic might be said to be that discipline which articulates and refines the standards (and their theoretical foundation) of right and wrong in matters of reasoning and argumentation. Broadly considered, the history of logic can be divided into two segments: the Aristotelian and the modern.

Aristotle is considered the founder of logic. His achievements are legion, most notably the first attempt at a conscious articulation of the standards of logic. In *Prior Analytics,* Aristotle developed the theory of syllogistic inference. In *Posterior Analytics,* he presented a theory of demonstration—reasoning designed to yield certainty. In *Topics,* he presented a theory of probable reasoning, while in *De Sophisticis Elenchis* Aristotle introduced what could be termed the theory of fallacy. The development of logic for the next 2,000 years shows the influence of his prodigious talent and labour.

It has been stated that no substantial development occurs in logic from

Aristotle until the time of Frege. While this broad statement should be challenged since it overlooks the important work of Mill, Boole, and others, it is serviceable insofar as it highlights the fact that Frege has come to be regarded as the founder of modern logic. His contributions to logic—most of them not recognized during his lifetime—are legion also. He presents, in his ***Begriffsschrift*** (1879) the first rigorous formulation of non-syllogistic logics: propositional logic and quantification theory. Frege also introduced the explicit distinction between axioms and rules of inference, thereby laying the foundation for the formal logistic systems of modern and contemporary logic.

There is no point in rehearsing here all the developments of the past 100 years in logic. What does require emphasis is simply this. When one speaks of the spectacular development of logic over the last 100 years, one is quite clearly referring to formal logic and its many relatives: semantics, pragmatics, metalogic, etc. In this progress, informal logic has not, so far, been a participant. Thus it is possible to say now about informal logic, the very same thing that might have been said one hundred years ago about formal logic: there has not been any significant development since Aristotle.

Perhaps this statement seems bold. But we direct attention to the fact that Kneale and Kneale's landmark history, ***The Development of Logic***, contains not a single mention of informal logic and scarcely any treatment of topics related to it. We are not suggesting that there is a *lacuna* in the Kneales' work. On the contrary, the point is that the conspicuous absence of treatment of informal logic in their work testifies to its undeveloped state.

In the last 25 years, however, there have been signs that the situation is changing and that informal logic has begun to take its place alongside formal logic as an independent branch of logic. We have more to say about these signs in Part II.

(B) Methodology. Before we outline the developments, we ought to say a few words about our methodology. If one is to survey recent developments in informal logic, it is clearly necessary to utilize some conception of what informal logic is. Simply put, our conception is that informal logic is that area of logic (not yet fully canonized as a discipline) which attempts to formulate the principles and standards of logic which are necessary for the evaluation of argumentation. We take this to include not only the development of procedures and techniques for appraising arguments but also the articulation of supporting theory.[1]

Using this as our working definition of informal logic, we reviewed the literature of the last 25 years. Our survey is divided into three categories: (A) monographs; (B) journal articles; and (C) textbooks.[2]

II. RECENT DEVELOPMENTS

Over the past decade something new has been emerging in logic. To call it a *Geist* is overblown, but suggestive. To call it an "outlook" is safe, but not forceful enough. The outlook we refer to is characterized by two interrelated features. First, there has been a turn in the direction of actual (i.e., real-life, ordinary, everyday) arguments in their native habitat of public discourse and persuasion, together with an attempt to deal with the problems that occur as a result of that focus. Second, there has been a growing disenchantment with the capacity of formal logic to provide standards of good reasoning that illuminate the argumentation of ordinary discourse. The result has been a number of initiatives to develop methods of identifying, analyzing and evaluating reasoning, which do not rely primarily on the instruments or nomenclature of formal logic. True, these initiatives have been sporadic, dispersed, and tentative. Yet they have also included some decisive forward thrusts. We believe, in short, that informal logic has begun to come into its own as an area of theoretical inquiry.

The important developments in this regard have occurred in the journal articles and in the burgeoning number of textbooks. We shall have a close look at the literature in these two categories after looking first, and rather more briefly, at monographs.

(A) Monographs. In our judgement, only three monographs of significance to informal logic have appeared in the last 25 years: Toulmin's ***The Uses of Argument*** [1][3], Perelman and Olbrechts-Tyteca's ***La Nouvelle Rhétorique*** [5], both of which came out in 1958 (although the latter was not translated into English, under the title, ***The New Rhetoric***, until 1969) and Hamblin's ***Fallacies*** [6], published in 1970. None of these monographs has had the impact it deserves in the philosophical world at large, nor even within the discipline of logic.

Those who have been working in informal logic will nod in recognition at what Perelman and Olbrechts-Tyteca were talking about when they wrote these words:

> Although it would scarcely occur to anyone to deny that the power of deliberation and argumentation is a distinctive sign of a reasonable being, the study of the methods of proof used to secure *adherence* has been completely neglected by logicians and epistemologists for the last three centuries. ([5], p.1)
>
> Logic underwent a brilliant development during the last century when, abandoning the old formulas, it set out to analyze the methods of proof effectively used by mathematicians. Modern formal logic became, in this way, the study of the methods of demonstration used in the mathematical sciences... Logicians owe it to themselves to complete the theory of demonstration obtained in this way by a theory of argumentation. ([5], p.10)

Clear though this clarion call was, it has been heard almost exclusively by rhetoricians—possibly because ***The New Rhetoric*** dwells more on the presentation of arguments than on their analysis or criticism.

Hamblin's and Toulmin's works have fared only slightly better. Toulmin's remarks about the problems he was going to consider echo those of Perelman and Olbrechts-Tyteca:

> ... they are problems which arise with special force not within the science of logic, but only when one withdraws oneself for a moment from the technical refinements of the subject and inquires what bearing the science and its discoveries have on anything outside itself— how they apply in practice, and what connections they have with the canons and methods we use when, in everyday life, we actually assess the soundness, strength and conclusiveness of arguments....
>
> ... the science of logic has throughout its history tended to develop in a direction leading away from these issues, away from practical questions about the manner in which we have occasion to handle and criticize arguments in different fields, and towards a condition of complete autonomy, in which logic becomes a theoretical study on its own, as free from all immediate practical concerns as is some branch of pure mathematics. ([1], pp. 1-2)

Toulmin proposes that rational assessment of arguments should be conceived on the model of judicial practice. One should look to see whether argumentation conforms to certain basic rules of procedure rather than on the model of geometric demonstrations. He argues that the component functions in argumentation are more plentiful and varied than merely the advancing of premises for conclusions, and urges further distinctions "between claims, data, warrants, modal qualifiers, conditions of rebuttal, statements about the applicability or inapplicability of warrants, and others." ([1], p.142) In the same spirit, Toulmin contrasts the "idealised" logic of symbolic logic with the "working" logic he thinks is needed for the analysis of everyday argumentation. Finally, he calls for a *rapprochement* between logic and epistemology; a broadening of logic to treat "arguments in all fields as of equal interest and propriety," and so compare and contrast "their structures without any suggesion that arguments in one field are 'superior' to those in another"; and a reintroduction of historical and even empirical considerations into logic. Despite these very interesting suggestions, if Toulmin's monograph has had much influence, it has gone largely unacknowledged.[4]

Hamblin's *Fallacies* has received the most widespread recognition of the three monographs. It is, for example, *de rigeur* these days to acknowledge his criticisms of the fallacy approach.[5] Where *Fallacies* has had its clearest influence is in the work found in recent journal articles, for it is there that the only work of a theoretical nature is being done on fallacies, and it was Hamblin who drew attention to the great need for such work. His monograph provides the only extensive history of writing about fallacies (an excellent one at that); it underscores the neglect that fallacies have been subjected to in logic texts, and by extension draws attention to the neglect of the whole of informal logic; and it offers a theory of fallacy of great interest, particularly because it builds from a concept of argument as used *in practice*.

(B) *Journal articles*. We begin with two prefatory notes. First, the underdevelopment of informal logic may be inferred from the fact that neither of the standard indices of journal articles (*The Philosopher's Index* and *Social Sciences and Humanities Index*) contains a separate heading or entry for informal logic. Second, because informal logic lacks a clearly established identity, it has no standard nomenclature. Hence, for example, the term 'practical reasoning' in the title of an article sometimes signifies that it is about informal logic ("practical reasoning" being one of informal logic's many *alter egos*), but often, of course, it indicates that the article belongs within ethics. So our survey of the journal literature was plagued by an identity problem, and no doubt we have erred by including articles that ought not to have been included, and by failing to include some which should have been.

1. Quantity.

Our survey provides evidence that interest in informal logic has escalated especially in the last 10 years. In the 15 years prior to 1968, there were only 9 articles that seemed pertinent to informal logic. Since 1968, we count some 58 articles about aspects of informal logic, most of which (40) have been published in the last four years.

2. Distribution.

Although articles on informal logic have been widely distributed in the philosophical journals, there is a noticeable concentration of them in *Philosophy and Rhetoric*. From 1971–77, 53 articles appeared, of which 16 (almost a third) appeared in that journal. This finding is not surprising, for there is no journal of informal logic (where such a concentration would be expected), and *Philosophy and Rhetoric* comes as close to being such a journal as any.

3. Principal Researchers.

The most prolific contributors have been Woods and Walton. Since they began to publish work in this area, in 1972, they have jointly published 12 articles—over a quarter of the total production (47) since that time.

4. Principal Areas of Research.

Journal articles have focussed on either of two areas: (a) the theory of fallacy, and (b) the theory of argument. Let us look at each of these in turn.

(a) By "theory of fallacy," we mean the attempt to formulate with clarity and rigour the conditions under which a particular fallacy occurs, along with related questions about the nature and/or existence and/or classification of various kinds of fallacy. It seems clear that research in this area was stimulated by Hamblin's work, and to a lesser degree by that of Perelman and Olbrechts-Tyteca. The charter for such research has been well formulated by Woods and Walton:

> We neglect the study of fallacies at our peril, for it is just in these areas that rational criteria, however inexact and tentative, are sorely needed as an aid to the adjudication of actual, everyday argumentation. While the traditional treatments of the fallacies are too unsystematic to be useful as an effective device in argumentation, their abandonment leaves a gap that no one (as yet) quite knows how to fill. Hamblin suggests that we are in the position of the medieval logicians before the 12th century. We have lost the doctrine of fallacy and need to

rediscover it. ([79], pp.17–18)

In their own work, Woods and Walton have attempted to fill this gap in the theory of fallacy by providing more rigourous treatments of these informal fallacies: *argumentum ad verecundiam* [52]; *ad baculum* [71]; *ad hominem* [79], [80]; *post hoc, ergo propter hoc* [81]; and *petitio principii* [61], [62] and [82].

On the whole, the informal fallacy that has most captivated the interest of researchers is *petitio principii*—or *begging the question*. The eleven articles devoted to this fallacy range widely in focus and approach.[6] Is there such a fallacy or not? Robinson [35] argues that there is not, although most other writers take the position that there is, but that its nature is not clear. A number of different approaches have been explored, summaries of which can be found in Woods and Walton [61] and in Sanford [77].

After *begging the question*, the fallacy that has been most discussed is *ad hominem*. The eight articles on that fallacy again display diverse approaches,[7] including the question of whether or not such a fallacy exists, raised by Gerber [48]. Most writers take it that there is such a fallacy but are not in agreement about its nature and types. Excellent discussions of the various positions are to be found in Woods and Walton [79], and in Barth and Martens [72].

The other informal fallacies treated in journal articles are: *composition and division* [18], [19], [25], and [55]; *the appeal to force* [59], [60], [63]; *the argument from authority* [52] and [54]; *many questions* [42]; *the appeal to ignorance* [36]; and arguments from *analogy* [51].

While research on the theory of fallacy has begun to fill the gap mentioned by Woods and Walton, a great deal of work remains. Of no informal fallacy can it be claimed that we now possess a widely accepted theoretical account, and many of the important informal fallacies have not yet been investigated in a theoretical way at all: e.g., *straw man*, and *two wrongs*. Indeed, by any standard, one of the most important informal fallacies is *irrelevant reason* (*"non sequitur"*), yet an adequate non-formal analysis of the concept of relevance has yet to be carried out. The attempts of Anderson and Belnap[8] (and their successors) to capture the notion of relevance in a formal system have not been entirely successful. Whether informal logic can fare better in this task, only time and further research will tell. Again, the concept of adequate or sufficient evidence, as it relates to everyday arguments, requires conceptual underpinning. And under what conditions is an undefended premise in an argument logically offensive? Vagueness is inherent in much mundane argumentation, but the concept of vagueness requires careful analysis if it is to be employed in effective logical criticism.[9]

In sum, although some interesting beginnings have been made in research on the theory of fallacy, this area of informal logic is still in its infancy.

(b) By "the theory of argument"—the second focus of research in the journals—we mean the attempt to formulate a clear notion of the nature of argument which is not beholden to formal logical or proof-theoretic models, and to develop principles of criticism and reasoning which come closer to shedding light on natural argumentation than do those of formal logic.

Again, research on this topic shows the impact of Hamblin's work, and here we quote from Woods and Walton's critical discussion of Hamblin in which they call for

> ... the eventual emergence of a concept of argument more adequate to the domain of natural argumentation and of informal fallacies than the purely syntactic proof theoretic accounts that, by themselves, are appropriate only to the domain of rigorous mathematical demonstration. ([91], p.104)

Woods and Walton have themselves contributed to the development of such a concept in the above article, and also in [70] and [83].

Other articles worth mentioning are these: Apostel [31] attempts to show that assertion logic (a branch of modal logic) is "urgently needed for the description of discussion and argumentation" (94) and claims further that "it is possible to build a bridge between descriptive and normative aspects of the theory of controversy by means of the concept of 'competent audience' " (107)—which concept he tries to elucidate.[10] Brockreide [37] examines the implications of the practice of argumentation by the introduction of a sexual model. Thus, some arguers are, according to the analogy, rapists; some, seducers; and some are lovers. Such an analogy may be a useful pedagogical device for explaining the nature and purposes of argumentation. Iseminger [49] argues that there is a plausible sense of the term 'successful argument' in which success consists in more than validity, but that the additional condition is not and does not entail the truth of the premises, and so success does not amount to soundness. Can Iseminger's concept of successful argument be developed as an alternative to soundness for the informal assessment of arguments? Can this concept be used to bypass the inductive-deductive hegemony on argumentation which informal logic must perhaps resist?[11] Kruger [57] presents a new system of classification of controversial statements, claiming that "the student of argumentation will become acquainted with concepts which, though seldom discussed in the textbooks of his discipline, are the *sine qua non* of effective argumentation" (138). Is the notion of effective argument another alternative to soundness as an ideal for mundane arguments? Finally, Peppinghaus [69] has devised an interesting set of principles (the autonomy of the addressee, active openness, the golden rule of argumentation) which introduces an entirely different classification of logical miscues.

Philosophers' contributions to the theory of argument have in a number of articles focussed on the special case of philosophical argumentation. Schouls [19] argues that philosophical positions involve presuppositions in such fashion that philosophical communication is possible only among philosophers who share one another's presuppositions—a claim disputed vigorously by King-Farlow [33] and Kodish [34]. Johnstone [29] holds that there is an honourific sense of *argumentum ad hominem* and that indeed most philosophical arguments must take this form. Brown [28] takes the position that there are no onus-assigning propositions of any sort, "only onus-assigning contexts or situations in which the disputants find themselves"

(81). Facione [65] discusses the role of counterexampling in philosophical arguments and shows that there are four levels of philosophical debate to which the use of counterexamples may lead (529). There are then a number of promising lines of investigation for the development of a theory of philosophical argumentation.

To summarize, we would say that the theory of argument is not much further along than the theory of fallacy. The notions of effective argument, successful argument, inadmissible argument are all of them inchoate, but may be seen as initiatives in the direction of exploring a notion of argument which is closer to the domain of natural argumentation, and which may outrun the notions of validity and soundness.[12]

(c) The other articles included in our survey are not easily categorized. Certainly, one of the factors crucial to any successful enterprise in the area of mundane argument is the notion of *context*, which is underscored in Anderson and Mortensen [17], and Coreliss [38]. Another important topic in the appraisal of natural arguments is that of finding and formulating missing premises or assumptions. This question is treated by Lee [43], who argues that assumption-seeking is a type of hypothetic inference from a given belief to its proximate grounding. Hypothetic inference, which is based on supposability, is not reducible to deduction (necessity) or induction (probability). If correct, Lee's position may further erode the grip of the deductive/inductive distinction, since assumption-seeking (we will argue shortly) is crucial to the success of the informal logic enterprise. Both Griffith [56] and Scriven [30] make strong cases for the limitations of formal logic. Scriven, for example, says:

> In difficult areas like practical logic, the trained philosophers are rarely to be found, preferring to build their own ivory towers. What is an assumption? No logic text (of the seventy or so I have on my shelves) has an answer that can survive five minutes' search of the stockpile of potential counter-examples. Why bother to distinguish inductive arguments from deductive when virtually every practical argument can be reconstructed with equal plausibility in either form? Can one comprehensively criticize an argument in itself without considering alternative arguments for the same conclusion and alternative conclusions from the same premises? A dozen or more questions arise as one begins the serious study of effective reasoning—all bypassed by the supreme irrelevance of formal logic, which has never been shown to have either content or skill carry-over to the practical, and probably not even to the philosophical domain. (902)

Walton and Woods [53] outline a number of ways in which informal logic connects with other disciplines.

Summary: The research done in the journal articles of the last 10 years, and particularly the last four, shows an upsurge of interest in informal logic, with the theory of fallacy and the theory of argument as twin foci. But, as we shall see, other important areas in need of research have yet to put in an appearance. And while the research that has been done marks a good beginning toward filling the gaps in theory mentioned at the start, a clear direction shaping research efforts and a clear application to the realm of practice have been missing.

(C) Textbooks. Turning to textbooks, we find that over the past 30 years the changes have been dramatic. Borrowing terminology from the language of computer talk, we can divide the introductory logic textbooks that have appeared since the Second World War into two "generations." The first generation texts by and large belong to either of two paradigms, typified by Beardsley's *Practical Logic* (1950) [88], on the one hand, and by Copi's *Introduction to Logic* (1953) [89], on the other. First generation texts continue to appear on the market today, but by the beginning of the 1970s a second generation of introductory texts had begun to appear. These were anticipated by Little, Wilson and Moore's *Applied Logic* (1955) [91] and Fearnside and Holther's *Fallacy: The Counterfeit of Argument* (1959) [92]; however this second wave of texts, devoted to informal logic as their main focus, began in earnest with Michalos's *Improving Your Reasoning* (1970) [109], Capaldi's *The Art of Deception* (1971) [110], and especially Kahane's *Logic and Contemporary Rhetoric* (1971) [111]. There followed, and continues to issue forth, a spate of textbooks in the area of informal logic which defies simple categorization and which provides the strongest evidence for our claim that a new outlook is abroad.

The developments represented by these second generation (informal logic) texts can best be appreciated in contrast to the first generation paradigms from which they are such a departure. Therefore we shall begin our discussion of the significance of recent informal logic texts by sketching the main features of the Copi and Beardsley first generation paradigms.

Copi's *Introduction to Logic* exhibits the classic features of what we shall call the "global approach." Its three parts are intended to touch on all the main areas of logic. Part One treats language, and includes chapters on the uses of language, informal fallacies, and definition. Part Two treats deduction, with chapters on categorical propositions, categorical syllogisms, arguments in ordinary language, symbolic logic, evaluating extended arguments and propositional functions. Part Three treats induction, with chapters on analogy and probable inference, causal connections, and science and hypothesis. Copi was not the first to use this pattern (cf., Max Black's *Critical Thinking* (1946)[84] and H. L. Searles's *Logic and Scientific Methods* (1948) [85], but by virtue of the sheer number of editions (in its 5th edition in 1978) and printings, Copi's text is the best known, and so it stands as the preeminent example of this approach to logic and to informal logic.

This structure is repeated almost identically time and time again in introductory logic textbooks (see, for example, Carney and Scheer [97], Kilgore [104], Baum [120], Manicas and Kruger [127] and Blumberg [123]. Other texts come close to the paradigm, but omit one or another of the parts and have slightly different emphases. For instance, Schipper and Schuh [93] drops part III and devotes more time to informal fallacies; Barker [99] drops part I; Terrell [103] drops part I. Or, similar ground is covered, with slight variations, in a different order. Witness here Michalos [107] and Kahane [108]. Of course each textbook writer is likely to maintain that his or her text is unique—and no doubt in some respects each is. Nevertheless, the global paradigm dominates the first generation post-war introductory logic texts.

From the point of view of informal logic, two features of the global paradigm are especially significant. The first is the assumption that the rules of deductive logic and the principles of induction and scientific method are central and essential to the logical appraisal of *all* argumentation, for *all* purposes. Arguments, according to this approach, are simply either deductive or inductive; bad arguments are invalid or unsound. The second striking feature of this paradigm is its perfunctory treatment of informal fallacies —which Hamblin has dubbed the "standard treatment" and castigated:

> as debased, wornout and dogmatic a treatment as could be imagined —incredibly tradition-bound, yet lacking in logic and in historical sense alike, and almost without connection to anything else in modern Logic at all. ([6], p.12)

Combined with the neglect of fallacies, and the focus on formal models of argument is an inattention to the possibility that the appraisal of arguments in their live, everyday settings may require alternative or supplementary canons of evaluation. We are not saying that these writers would deny this possibility. The point is that their interests and sympathies in these texts lie elsewhere than informal logic.

The second paradigm exhibited in first generation textbooks we call the "critical thinking" approach. It looks like the real ancestor of the texts that are emerging in the 1970's. Beardsley's **Practical Logic** [88], for instance, covers a great deal of the same ground as does Scriven's **Reasoning** [129], written 26 years later. The critical thinking approach combines several features. Its focus is on practical skills in clear thinking that are applicable directly to one's functioning as a reasoning person in the various roles of everyday life. The tools of logic are employed as an adjunct to this objective and therefore are not presented as objects of study in and for themselves. More attention is devoted to meaning and natural language than to formal systems. Finally, the practical application of the skills taught is acknowledged by making the invented examples realistic and including in the exercises some actual arguments.

Although marked similarity exists between such early post-war critical thinking texts as Beardsley's and the more recent informal logic texts of the 1970s, we are inclined to regard the former as belonging to the earlier generation. Beardsley was still quite satisfied with sentential and predicate logic as useful tools for the analysis and evaluation of natural arguments, whereas more recent texts are either uneasy or unhappy with that assumption. Also, his invented examples have the order and elegance of a well-turned mind, and his borrowed examples come primarily from the sort of well-ordered reasoning typical of academic literature. More recent texts turn to the problems of grappling with the sort of poorly organized argumentation typical of popular contexts. Finally, second generation texts appear not to be written in a conscious return to the earlier critical thinking tradition, but instead as a conscious reaction against the hegemony of formal logic represented by the global paradigm.

Let us then turn to a closer examination of the second-generation or "New Wave" textbooks (as we shall sometimes refer to them).

We have examined 54 introductory texts published since the war that devote at least some space to informal logic, including a few that may not have been aimed directly at the textbook market when they were first published. The textbooks are identifiable by prefatory notes to students and/or teachers, exercises, or other internal evidence. These are texts written for a first course in logic, reasoning or critical thinking. We allowed "informal logic" to cover discussions of language (such as emotive terms, or ambiguity), of definition, of informal fallacies, of polls and statistics, and generally of any fare related to argument or reasoning outside syllogistic or formal deductive logic, or inductive reasoning and scientific method.

Of the total of 54, 25 textbooks came out between 1946 and our arbitrary cut-off date prior to the present decade, 1969; and 29 appeared since 1969.

What the figures show, when combined with our interpretive categories, is that 16 of the 25[13] pre-1969 texts followed the global paradigm exemplified by Copi [89], while five[14] fitted the critical thinking paradigm typified by Beardsley [88]. Each of the four remaining is *sui generis*.[15] In sum, there are twice the number of texts in the global paradigm as in the critical thinking paradigm, or some variant of it, in the first 25 years of our sample.

The picture has changed dramatically in the last ten years. Of 29 texts, only 11 belonged to the global paradigm,[16] and a number of them incorporate features of the New Wave texts. Another 15 belong exclusively to informal logic—three being devoted primarily to fallacies,[17] and 12 including other nonformal material instead, or as well.[18] The remaining three are attempts to combine the new informal approach with material on deductive and inductive logic.[19] So if our sample is any indication, there certainly has been a turn to informal logic in introductory textbooks in the last decade.

What is striking about the recent textbooks dealing primarily or exclusively with informal logic is not just their numbers, but much more interestingly, the new turns they have been giving to their subject matter. When we speak of a "New Wave" in informal logic, we have these changes in content and treatment primarily in mind. We have counted five significant developments. Together with the work in the journals, these are the bases for our claim that something important is happening in informal logic today.

1. Working with "natural" arguments.[20]

What could be more obvious than to put the analytic and evaluative tools of informal logic—for example, those of the fallacy approach—to work on arguments that have actually been used to try to persuade people, the sorts of arguments the student will encounter outside the classroom? Surprisingly, the analysis of such examples is rare, or nonexistent, in the first-generation texts that devote space to informal logic. Almost without exception, the examples are inventions by the author (or borrowed from other texts). Moreover, the examples are usually artificial. That is, they are simplified, clear, unambiguous; their premise-conclusion structure is evident; each statement plays a role in the argument. When natural arguments were used as illustrations or for exercise examples, they tended to originate from philosophical, other scholarly, or literary, sources.

The examples found in the 1970s texts tend to differ in three ways. More and more are natural arguments. When invented, their artificiality is minimal: their subject matter is topical, and their literary form is closer to the way ordinary people talk and write. And their source is the everyday public realm—newspapers and magazines, "popular" books—rather than literary, scholarly or philosophical texts.

This small change in the kind of examples used makes an enormous difference in both the theory and practice of informal logic. For such examples are rambling, confused, digressive, prolix discursions which are briar patches for the logician who would trace their logical flow and assess their strengths and weaknesses.

The primary result of dealing with actual examples is that the writer is forced to abandon preconceptions and face the actual data—to become more empirical in that sense. Any lack of fit between traditional categories (principles, distinctions) and everyday reasoning becomes dramatically evident when one tries to apply these categories to actual examples. Scriven has noted a case in point:

> The use of any calculus to handle problems that surface in reality (in natural language) involves ... *encoding* the original problem into its formalized representation.... [T]he problem with formal logic is that the encoding step ... is just about as debatable (in anything but trivial arguments where there's no need to use the calculus) as the assessment of the original argument. ([129], p.xv.)

What has happened in recent work, as a result of attention to actual examples, is what Toulmin [1] called for back in 1958: informal logic has become less *a priori* and more pragmatic.

Working with such actual, everyday persuasive discourse, the logician faces new problems. What *is* the argument? How is it to be extracted from its surrounding rhetoric? What verbal or contextual clues may be used, and how? What principles of interpretation apply? How is the argument to be displayed in order to exhibit its logical structure fairly and perspicuously? What standards of evaluation are then to be applied? How are the criteria of evaluation to be determined? In their practice—working with rather immediate pedagogical goals before them—recent informal logic textbook writers have struggled with and offered answers to these and other questions. The answers have not been uniform; indeed, the questions have not all been perceived by everyone, nor perceived in the same ways. What is needed now, in fact, is a survey of the various practical solutions to these problems that have been developed and an attempt to fashion the necessary theoretical underpinnings.

We cannot leave this point without giving credit to Kahane for his major contribution to the breakthrough into everyday argumentation. As far as we know he was the first writer to use everyday examples almost without exception throughout the body of his text as well as in the exercises. Since the example of *Logic and Contemporary Rhetoric* [111] in 1971, there has been no excuse for manufacturing silly, artificial examples, or for fashioning exercises to fit neatly into a writer's *a priori* principles of evaluation.

2. The treatment of fallacies.

We have quoted Hamblin's now-famous castigation of The Standard Treatment of fallacies, which applies particularly to first generation texts taking the global approach. Hamblin shows that there has been no one single tradition in the treatment of fallacies. Although, it was part of the conventional wisdom predating his book that there is no given principle for the individuation or classification of fallacies, first generation texts tended to classify informal fallacies either according to Aristotle's division (into language-dependent or "material" fallacies and other-than-language-dependent, sometimes called "psychological" fallacies) or some variant of Aristotle's division; or else they classified fallacies as deductive or inductive. These texts stocked their informal logic inventories with fallacies that had become established fare by the 19th century. In other words past practice continued of its own inertia.

That has changed in New Wave texts. There is innovation in the selections of fallacies treated, in their classification, and in the formulation of "new" fallacies. These texts tend to use fallacies as tools for the teaching of practical skills in critical thinking rather than to discuss them out of a sense of an obligation to expose students to the traditions of the past.

The primary criterion of fallacy classification in recent texts seems to be pedagogical. Michalos's *Improving Your Reasoning* [109] is an interesting example of the change. Michalos classifies fallacies as formal and informal, and divides the latter into those occurring when the argument is deductively or inductively valid but has false premises, and those with irrelevant premises. But after this nod to conventional wisdom, he turns to his declared objective, that of helping students improve their reasoning, and proceeds to group his fallacies on entirely different, and heuristic, grounds: question-begging fallacies, fallacies of pseudo-authority, confusion, political fallacies, and so on. Fearnside and Holther's *Fallacy: The Counterfeit of Argument* [92] is similarly transitional. It uses the major Aristotelian divisions, but then introduces eight sub-groupings, under headings such as "stirring up prejudice," "rationalization and lip service" and "diversions" that are clearly pedagogical in inspiration.

Examples of pedagogically-motivated classifications abound. Capaldi [110] groups fallacies according to whether they are more likely to occur when one is presenting a case, or attacking an opponent, or defending a case. Then, because of their separate importance, he devotes individual groupings to political propaganda and cause and effect reasoning. Kahane's [111] fallacy groups—"fallacious because invalid" and "fallacious even if valid"—cut across the deductive/inductive distinction, and ignore the Aristotelian division. He also includes a chapter on statistical fallacies. Engel [125] distinguishes fallacies of ambiguity, fallacies of presumption and fallacies of relevance; and within the second category he subdivides further: fallacies overlooking the facts, evading the facts and distorting the facts. Ehninger [117] classifies into three different groups: fallacies of language, of thought, and of tone and manner. Johnson and Blair [132] divide their list into five sections: fallacies of diversion, of impersonation, of sleight of hand, of prejudgement and of intimidation. Fogelin [133] uses just two categories: clarity and relevance.

The actual lists of fallacies have changed in at least five ways. (a) By and large, Latin labels have been translated in English, or replaced with more descriptive English labels. This is a minor point, but it is a symptom of the release from tradition and the new practical preoccupations. (b) Distinctions—particularly the one between equivocation, amphiboly and accent—have been collapsed. The point seems not to be to cover all the theoretically possible subdivisions, but to grasp the central idea that can illuminate concrete assessment of arguments. (c) A number of fallacies found on the first-generation lists have been dropped. For example, one sees less and less of "appeal to force" or "accident"; and "composition" and "division" are rarely found in New Wave texts. Why? In some cases the excised fallacies belonged to a now-defunct tradition of debate. Others seem to have been restricted to academic or artificial contexts. (d) Quite often the standard treatment of first generation texts has been expanded. "False cause," for example, can be the occasion for an extended discussion of ordinary causal reasoning. "Complex question" is the subject of entire sections on the use of assumptions in everyday reasoning. "Appeal to authority" has given rise to discussions of knowledge and belief, and of kinds and qualifications of authorities. (e) Finally, completely new fallacies have been added as writers have canvassed real arguments about current issues that appeal to present day beliefs and attitudes, and have pondered the responsibilities of arguer and audience as reflective citizens or consumers. We think here of Kahane's [111] "tokenism," "unknown fact," "suppressed evidence" and statistical fallacies; or of Weddle's [136] "stereotyping" and "half truth."

Finally, we should note that fallacies are no longer always assembled in a single chapter given over to their brief exposition—except in texts within the global paradigm. Now several chapters, even entire books, are devoted to their exposition and exemplification. Or else references to fallacies are scattered throughout the text, invoked when and as the writer deems it useful. The guiding principle seems to be that the fallacy approach should be used as an adjunct to teaching reasoning skills, and incorporated into a textbook on informal logic on that basis.

From a theoretical point of view, of course, the treatment of fallacies in New Wave textbooks seems nothing short of chaotic. Definitions of the concept of fallacy vary, classificatory schemes abound, the treatments of individual fallacies have little uniformity, and there seems to be no real principle of collection.[21]

The recent treatments of fallacies might, so far as the theory of fallacy goes, be considered a shambles. However—and this is the present point—there have been some extremely striking practical innovations in these texts. What is needed is the generation of theory out of practice. And the writers who have been doing the theoretical work in the journals should take the New Wave textbook handling of fallacies seriously, to the point of considering it raw material or data that cannot be ignored.

3. *Consideration of "full" or "extended" arguments.*

For the purpose of exemplifying fallacies, writers belonging to the earlier tradition customarily relied on short passages which, as we have been pointing out, they usually had to invent. More recently, as a consequence of turning their attention to natural arguments, informal logicians have been forced to reckon with fully-developed or complete arguments. In everyday situations, people usually try to give as much support for the claims they are advancing as they can, or as they think appropriate on the occasion, and this usually means developing a fairly full case for those claims, not just single, one or two step inferences. In trying to develop textbooks that would help their students to interact with such extended arguments, New Wave writers have themselves had to grapple with the practical questions of how to analyse and assess them. The results have been uneven, but on the whole we think they have been the most exciting and significant of the New Wave text developments.

The strategies recommended to students differ from text to text (cf. Angell [96], Kahane [111], Thomas [115], Ehninger [117], Scriven [129], Johnson and Blair [132] and Fogelin [133], to mention some we have canvassed), but they invariably cover two major tasks: (a) the interpretation (including the extraction and structuring) of the argument, and (b) the evaluation of the argument. In dealing with the first of these, New Wave writers have had to answer three interrelated questions. First, what belongs and what does not belong to the argument? Second, what should be added—supplied by the critic—to complete the argument? Third, what is the structure of the argument? What is the organization of its logical flow from support to conclusion? How can this structure be most perspicuously displayed? An answer to any one of these affects the answers to the other two. The textbook writers have been trying to come up with practical guidelines for their students in each of these areas. The second major task, that of criticism, is similarly complex, and dealing with it has been the occasion for additional advice in the textbooks.

The remainder of this section undertakes to describe in greater detail the results of grappling with these two questions, under four headings: (a) the conceptualization of argument structures, (b) the handling of missing premises and conclusions, (c) the ethics of interpretation and evaluation, and (d) the criteria of evaluation.

(a) The conceptualization of argument structures. New Wave authors have been trying to understand and set forth the different kinds of premise-conclusion relationships that are found in natural argumentation. The issue here is not the logical character of the reasoning or inferences; i.e., whether they are deductive or inductive.[22] Instead, what people are trying to map are the various kinds of arrangement possible in the way that premises support a conclusion. For example, Thomas [115], following Beardsley [88], sorts arguments into two types: convergent (where the premises work together to support the conclusion) and divergent (where the premises individually are supposed to support the conclusion). Ehninger [117] has adopted Toulmin's evidence-warrant-claim method of analyzing structure.[23] Scriven [129] has pointed out that an arguer may acknowledge deficiencies in his case, evidence that weakens the support provided by other premises, and so suggests a tree structure marking positive and negative reasons. Most writers have noted that premises themselves can be supported in extended arguments, and that indeed there often exist several levels of support. While many different methods of mapping the structure of arguments have been explored, questions remain. Should the structure show distinctions between major and ancillary support for a conclusion? Can all argument structures be catalogued? What is the most perspicuous way to display the structure of an argument? Is it pedagogically useful to expose structure in such complicated ways as some do (i.e. Johnson and Blair's [132] standardizing procedure)? Or would it be more effective to follow Kahane [111] and simply summarize the argument in outline form? Or to follow Fogelin [133] and simply note the argumentative functions (such as hedging terms, slanting, discounting, etc.)?

Once listed in this way, these questions strike one as perfectly straightforward. An outsider might register some surprise that more progress hasn't been made in providing answers and that there is not a wider range of more thoroughly worked-out alternatives. What needs to be pointed out, however, is that these question were not even *asked* until recently! New Wave texts deserve credit not for having provided the necessary conceptualizations, but merely for having recognized the need for them.

(b) Supplying missing premises and conclusions. Natural arguments are usually incomplete. They make leaps from supporting reasons to claims based on them that would be plausible only if certain other assertions, which they do not mention, were also accepted. Or they list reasons which, given all the contextual signals, are supposed to lead one to accept some claim, but they don't state that claim.

Anyone who has tried to evaluate natural arguments will know that these missing premises and conclusions must be formulated, for the strength or weakness of the argument very often depends on what they are. Anyone who has tried to formulate them or theorized about how to do this will know what a tricky job that can be. Should a missing premise be trivial, or lend substance to the argument? If the latter, how strong, or weak, should it be? On what grounds is one to answer these questions? The astounding thing is that the intricacies of formulating missing premises have just not been recognized and addressed.[24]

If recognizing a question is halfway to its answer, we would give half credit to some (but only some) of the New Wave informal logic texts for answering the questions about missing premises. Some who have discussed argument analysis have not even acknowledged the whole issue of unexpressed premises (e.g., Munson [128] and Fogelin [133]). Others have recognized the phenomenon, but gone no further (e.g., Thomas [115] and Ehninger [117]; or else have offered only brief suggestions about how to formulate these premises (e.g., Annis [116] and Johnson and Blair [132]). We have found only two texts that give *detailed* consideration to the problem of identifying missing premises and conclusions: Angell's **Reasoning and Logic** [96] and Scriven's **Reasoning** [129].[25] On the other hand, what both Angell and Scriven have to say is very good, and we would like to replay it briefly.

Although the two writers are in essential agreement, Scriven does not appear to have known of Angell's earlier treatment. Between them, they differentiate the missing premise that needs to be formulated for the purpose of argument assessment from, on the one hand, unstated reasons that may underlie various parts of the argument, but that lie "outside" the argument (Angell, 384), and, on the other hand, assumptions that are stated within the argument, but are identified as assumptions (Scriven, 81). Scriven characterizes the quarry as "the further assumptions that are required, in many cases, to make an inference satisfactory" (81); Angell speaks of "reasons . . . omitted from an argument . . . [that are] essential to its structure as it is stated" (384). Both note that the skill in finding missing premises calls for imagination (i.e., that there is no algorithm), and both offer practical guiding principles. Here is Angell:

> Though the methods proposed do involve certain assumptions and may require the exercise of imagination to some degree, the assumptions can be justified rationally and the imagination required is a disciplined one guided by certain principles and logical clues. (386)

Here is Scriven:

> Exactly how does one correctly formulate the missing premises of an argument? Here again, we find that imagination and originality are often required in this basic part of the critical process. (85)

Scriven goes on to state and defend three criteria that formulations of missing premises should meet. He and Angell agree about the first two of these—first, that "assumptions have to be strong enough to make the argument sound" and second, that "they should be no stronger than they have to be " (Scriven, 85; Angell, 386–387). They disagree about the third: Scriven argues that "you also want to try to relate the assumptions as you formulate them to what the arguer would be likely to know or would believe to be true" (85), whereas Angell argues, "we do not care whether the person who first presented the argument had these reasons in mind or not" (387).

Also, Scriven takes a stronger line against trivial missing premises. Angell allows that it is "sometimes satisfactory, simply to form the missing premiss by putting the reason *p* and the conclusion *q* into a conditional 'If *p* then *q*,' " but asserts that in many cases where this is done the results fail to have much plausibility (388). Scriven is firmer. He says, "It is entirely unhelpful to point out that a particular argument 'assumes' that its premises imply its conclusion" (163). And he provides a useful discussion of how to distinguish between what he calls "significant" and "insignificant" assumptions, together with supporting arguments (162-166). Our contention is that *any* further thinking about missing premises or assumptions ought to begin with a careful look at Angell and Scriven.

(c) The ethics of argument presentation, interpretation and criticism. Everyday arguments are digressive, rhetorical, repetitive, ill-organized, incomplete, and multi-functional. Trying to provide guides for analyzing and evaluating such arguments leads very quickly to such questions as: "Is it fair to treat that comment as part of the argument?" or "Ought the missing premise be framed to commit the arguer to so strong an assumption?" or "Should the arguer be castigated for every slip, no matter how minor?" In other words, we have to have an ethics of argument analysis and assessment.

Here again is a point which, once observed, seems to be obvious enough, but which surprisingly was not even noted until New Wave texts began to appear. In this case, too, the encounter with full-sized, living arguments seems to have functioned as the catalyst. What has emerged in answer to these questions is something that has been widely dubbed The Principle of Charity—the basic idea behind it being that one should give the arguer the benefit of the doubt.

Various principles of charity have been proposed—notably by Thomas [115], Baum [120], and Scriven [129]; and the principle has been used, though not named, elsewhere (e.g., Johnson and Blair [132]). We cannot state *the* Principle of Charity, not just because it comes in different formulations, but also because we have found it to function in at least four different areas. First, it is used in locating arguments. Thomas uses it this way when he says, "if a passage contains no inference indicators or other explicit signs of reasoning *and* the only possible argument(s) you can locate in it would involve obviously bad reasoning, then categorize the discourse as *non*argument" (9). Second, having decided that an argument is present, the principle is used to identify the content of the argument, e.g., by Scriven: "Be sure that wherever there's a vague sentence ... you cross it out ... and rewrite it in a more precise and perhaps more charitable form" (76-77). Third, the principle is used in formulating missing premises. Here is Baum's statement of it in this context: "When supplying missing premise or conclusion statements ... one should adhere to the *Principle of Charity*, which stipulates that one should supply statements that make the argument as good as possible" (135). Fourth, charity is recommended in criticizing arguments once they are located, identified and filled out. Here is Scriven on this point:

What the Principle of Charity does mean is that "taking cheap shots" is something we shouldn't waste much time doing. Other words that come in from ordinary language about this point are "nit-picking" and "attacking (or setting up) a straw man." These terms all refer to poor argument analysis, either to making irrelevant criticisms or to making criticisms that are not relevant to the main thrust of the argument or that are unfair in some other way. (71–72)

Although we have not given a full account of these writers' various principles of charity or their rationale in support of them, it is clear that questions abound. What is the rationale for a Principle of Charity? Is its justification in one context transferrable to another? How is it (or are they) to be formulated? Are there exceptions? Are there other, conflicting principles of interpretation or criticism?

The Principle of Charity is not the only ethical principle that has been raised in connection with argumentation. Looking at it from the side of the person who gives the argument instead of the critic, Flew [121] has proposed that arguers too have ethical obligations. He says, "to the extent that I make claims to knowledge without ensuring that I am indeed in a position to know, I must prejudice my claims both to sincerity and ingenuousness" (115). So, if Flew is right, to advance reasons in support of a conclusion is to take responsibility for the acceptability of those reasons.

Once again (it is becoming a refrain) New Wave authors have raised the issues and made plausible suggestions about their resolution, and opened up a topic for fruitful theoretical analysis.

(d) Standards of evaluation. The final spinoff from examining full-sized arguments in their original settings is a new perspective on argument criticism. Most texts still talk simply in terms of validity, soundness or fallaciousness. Arguments are conceived as good—or bad. Such unqualified judgements are too simplistic to be significant or interesting verdicts about most everyday argumentation. Reflective response to reasoning is fuller, more detailed, and balanced, like this: "There is something to what you are saying, but you should not rely so heavily on that one report." Or: "You have missed one of the strongest reasons for your position." Or again: "You do indulge in some mudslinging, but it is certainly hard to challenge your two central points."

There are signs in some of the New Wave texts of the evolution of more perceptive canons of criticism. Munson [128], for example, suggests that the adequacy of reasons is a matter of degree; he introduces the critical category of the *fairness* of the premises; and he sees the critical process as leaving room for reply and revision (187–197). At least by implication, Thomas's [115] "weak," "moderate" and "strong" classification of argument strength invites corresponding degrees in critical assessment (69–79). Johnson and Blair [132] have tried to introduce the ideas of degrees of critical strength and the opportunity for revision into the fallacy approach. For example, they rank fallacy charges from strong (irrelevance), through intermediate (insufficient evidence), to weak (disputable premises) (29). The most perceptive and imaginative suggestions about argument evaluation have been made by

Scriven [129]. After formulating The Principle of Charity as it applies to argument criticism ("no cheap shots"), he breaks the assessment process into three steps (43–45). First, criticize inferences and the premises—and in doing so, discriminate between the main conclusions and their support, and focus on the key weaknesses. Second, consider other relevant arguments, in order to put the strengths and weaknesses of the argument under scrutiny into perspective. Third, go back over your criticisms, considering their potency, and give the argument an overall evaluation. Try to judge, all things considered, how good or bad the argument is. After reflecting on such a rich critical strategy, how restrictive the tradition of working exclusively with validity, soundness and fallaciousness seems!

So far we have noted that New Wave textbooks have been noteworthy in three respects: (1) in the use of actual examples, (2) in the treatment of fallacies, and (3) in the development of strategies for the analysis and evaluation of extended arguments. The dominant themes are the growing independence from *a priori* theory and the pedagogical focus. These two themes are exhibited in two additional features of these textbooks which we should mention before leaving this section of the paper.

4. *The partial abandonment of the deductive-inductive dichotomy.*

Any break from this orthodox doctrine is far from complete; nor is there so much a denial of the distinction as there is a rejection of its usefulness for the appraisal of arguments in most public discourse.

Kahane [111] was one of the first to underplay the distinction. He observes:

> this standard division is not very useful ... it is rare in daily life to claim deductive certitude for the conclusion of an argument. (32, 2nd edition)

Thomas [115] went a step further in proposing that validity is a matter of degree, with deductive arguments only "at the highest end of the spectrum" having the truth of their conclusions "100 percent guaranteed" (75). He claimed, further:

> Empirical study of undoctored examples of reasoning in natural language seems clearly to show that in different arguments, the reasons lend different degrees of support to the conclusion. (72–73)

Scriven [129] takes a similar position. In a section where he contrasts an inference relying on the laws of arithmetic and one "made probable" by the premises, he states:

> That [—the latter—] type of argument is sometimes called an "inductive" argument, by contrast with the "deductive" one given earlier. ... The difference is not really very important from the point of view of practical reasoning because exactly the same choices are open to the respondent. The opponent must rebut either the premises or the chain of reasoning that takes us from those premises to the suggested conclusion. (33–34)

The point seems to be widely accepted in New Wave texts: analysis and criticism of argumentation that is worthwhile from a practical point of view cannot be viewed any longer as a minor subdivision of formal logic, and indeed it is time that it be incorporated as (at the very least) a semi-autonomous enterprise.

5. The widening scope of informal logic.

The scope of informal logic has been widening over the past decade, and neglect of the deductive-inductive paradigm is just one sign of movement out from under the wing of formal logic. Here are some others.

First, there have been certain shifts in emphasis even within the traditional territory of language, fallacies and definition. We have already discussed changes in the treatment of informal fallacies. In addition, recent work in the philosophy of language has begun to filter into informal logic textbooks. Fogelin [133] is a striking example, with its sections on speech acts, performatives, conversational implication and levels of language. The standard chapter on definition has been reduced in size or its components have been scattered throughout the texts to places where they become strictly pertinent to argument analysis (cf. Ehninger [117] and Scriven [129]). Moreover, the entire approach to definition has acquired a flexibility well illustrated by an excerpt from Weddle [136]:

> If the important thing [about definition] is getting the meaning across —that is, teaching the term's correct use—then any consideration of form and method should be judged by its ability to achieve that end. Thus, a satisfactory definition might take the form of . . . a Bronx cheer (as in defining "Bronx cheer"). (62)

The second point is that the traditional boundaries of informal logic have been extended. The analysis and evaluation of extended arguments is one sort of extension. The techniques of argument extraction and display, and of evaluation, while requiring attention to meaning and logical error, take them onto new ground. On another front, the recognition that it is necessary to have full and accurate information in order to assess everyday reasoning has led writers to appraise various *sources* of information. Many have included sections on polling and statistics. Gordon [100] was a pioneer in combining a treatment of fallacies with a factual and critical study of news media. Kahane [111] went further and discussed advertising and textbooks as well as news. Johnson and Blair [132] follow Kahane. Thomas [115] takes up decision-making. Fogelin [133] devotes close to half of his text to specimens of arguments in the domain of public policy, law, morality, theology, science and philosophy. In short, informal logic is increasingly seen as the tool for the critical analysis of reasoning, and its raw material, *wherever* they occur.

Summary of Part II. Few monographs have been written on informal logic. While especially those of Toulmin, Perelman/Olbrechts-Tyteca and Hamblin are in our view important works, they have had little influence upon work appearing in journals and textbooks—with the exception that Hamblin is widely mentioned.

The last ten years especially have seen marked growth in numbers of journal articles and textbooks in informal logic. The work in the journals has mainly been theoretical. The textbooks have been at the introductory level, and concerned with the practicalities of teaching useful skills to non-specialists. The textbooks—particularly those of the New Wave—have introduced innovations which have theoretical implications, but those theoretical issues have not been explored. In sum, there has been no significant interplay between the theoretical work of the journals and the innovations in practice found in the textbooks.

III. CONCLUSION

This chapter has been an interpretive report of the developments in informal logic over the past decade. The gist of our findings is that informal logic is in an adolescent stage. Not yet an independent discipline within logic with a clear and distinct identity, it nonetheless has shown enough growth and development to warrant attention in its own right. This growth has occurred in spurts, without theoretical coherence. Those working in informal logic have an increasing confidence that this is a separate field, even while they "try on" different topics to try to ascertain its parameters. The theoretical accomplishments in the field are spotty, and for this and other reasons informal logic has yet to attain respectability in the eyes of philosophers and logicians, especially those who know little about it.

What is clearly needed in this emerging field is a sense of definition and of direction—so long as the search for these does not distract from or paralyze ongoing research. We suggest the following as conditions of further development:

1. Informal logic needs to develop an even better understanding of what has already been achieved. This chapter is a beginning in this regard.

2. Informal logic needs to develop a clearer conception of its own identity and nature, of its component parts, of its scope and its relationship to cognate inquiries in logic (semantics, pragmatics, formal logic) and philosophy (epistemology and the philosophy of language), and to other disciplines (rhetoric, communication studies, debate, etc.).

3. Informal logic must generate an overview of the major issues which confront it and the major problems that require solution—along with the methods available to handle them. [Cf. Appendix.]

4. Finally, informal logic must recognize and deal with the obstacles to further growth. There seem to us to be two principal obstacles:
  a) the absence of any journal of informal logic;[26]
  b) the gulf between theory and practice. Certainly the existence of a journal devoted explicitly to the aims and to the advancement of informal logic would help to cure the gulf between theory and practice.

We end this chapter, then, simply with this list of tasks that lie ahead.

Appendix

An unclassified and partial list of problems and issues in informal logic.

1. *The theory of logical criticism:*
What is the purpose of logical criticism? Can an overall theory of logical criticism be developed? What are the criteria to be invoked in logical criticism?

2. *The theory of argument:*
What is the nature of argument? How is it related to reasoning? Is there a value to developing a typology of arguments? What are the standards that arguments (particularly mundane arguments) should meet? What principles should be decisive here?

3. *The theory of fallacy:*
What is the nature of fallacy? Can the conditions of individual fallacies be identified? Can fallacies be individuated? How should fallacies be classified? Is there a correct principle of fallacy classification? Should the notion of fallacy be junked?

4. *The fallacy approach vs. the critical thinking approach:*
What are the merits, and drawbacks, of each? Should/can they be integrated? Is this a pedagogical question only?

5. *The viability of the inductive/deductive dichotomy:*
Are mundane arguments one or the other? Are the validity/soundness criteria of evaluation inappropriate or outmoded? If so, what should replace them: effective argument? successful argument? plausible argument? persuasive argument? what?

6. *The ethics of argumentation and logical criticism:*
Can principles be formulated that assign the responsibilities of give-and-take in argumentation? What is (or are) the Principle(s) of Charity? What is their best formulation? What is their justification? Are there false principles of charity—or principles of false charity? Are there other, perhaps conflicting, ethical principles that apply?

7. *The problem of assumptions and missing premises:*
What exactly is a missing premise? What different kinds of assumptions can be distinguished in argumentation? Which are significant for argument evaluation? How are missing premises to be identified and formulated? Are these just practical and pedagogical questions, or theoretical as well?

8. *The problem of context:*
How does the context of argumentation affect its meaning and interpretation? What are the significant components of that context? Is a theory of contextual or pragmatic implication required for logical criticism?

9. *Methods of extracting arguments from context:*
How do principles of evaluation apply here? Is some theory of argument or reasoning necessarily presupposed? To what extent are the issues pedagogical and to what extent theoretical? Are there alternative but equally viable methods of extracting argument?

10. *Methods of displaying arguments:*
   Is there any evidence that some kinds of display (or presentation of structure) are more efficacious than others? What criteria can be invoked to adjudicate between various methods?

11. *The problem of pedagogy:*
   What alternative pedagogies are there for teaching informal logic? Are there criteria for adjudicating between them?

12. *The nature, division and scope of informal logic:*
   What is informal logic? What are its component parts or subdivisions? What should be included in a map or outline of its geography? On what basis if any can it be determined that the criticism of news media and advertising lie within the scope of informal logic? Is decision-making an area within the scope of informal logic? Are there other as yet unspecified topics that lie within its scope?

13. *The relationship of informal logic to other inquiries:*
   How is informal logic related to formal logic, semantics, pragmatics? How is informal logic related to other areas of philosophy, such as epistemology and the philosophy of language? How is informal logic related to such other disciplines as rhetoric, the theory of debate, communication studies, the psychology of reasoning?

Footnotes

[1]This is hardly a precise definition. However, it is our judgement that an attempt to produce a tight specification of informal logic, at this early point in its development, would be premature. Readers wanting amplification about the domain of informal logic should consult the Introduction.

[2]We have restricted attention to work done by philosophers and logicians in the English-speaking world. This means that we have not made any attempt to explore the connections between informal logic and other cognate disciplines such as rhetoric, the tradition of debate, pragmatics, semantics, etc.

[3]The numbers in square brackets following a book or article title, or an author's name, refer to the location of the complete bibliographical information about the work so indicated, which is contained in the authors' "A Bibliography of Recent work in Informal Logic," which is found at the end of this volume.

[4]Angell [96] (1964) lists *The Uses of Argument* in a chapter-end bibliography, and Ehninger [117] (1974)—a professor of rhetoric, by the way—employs Toulmin's evidence-warrant-claim distinctions for structuring arguments.

[5]Cf., for example, Copi [89] 5th edition, p. 87.

[6]Johnson [23], Williams [26], Hoffman [32], Robinson [35], Sanford [39], [77], Woods and Walton [61], [62], [82], Barker [64], Biro [73].

[7]Reipe [24], Johnstone [29], Finocchiaro [47], Gerber [48], [75], Barth and Martens [72], Woods and Walton [79], [80].

[8]Cf. for example, their "The Pure Calculus of Entailment," *Journal of Symbolic Logic*, 21: 19–52 (March, 1962).

[9]Cf. Machina [68] for an attempt to elucidate this concept using the machinery of formal logic.

[10]In our judgement, the payoff and degree of illumination which can be expected by using the conceptual apparatus of formal logic remains an open question. Here we can refer to the words of Bar-Hillel: "I challenge anybody here to show me a serious piece of argumentation in natural language that has been successfully evaluated as to its validity with the help of formal logic . . . The customary applications are often careless, rough and unprincipled, or rely on reformulations of the original linguistic entities under discussion into different ones . . . through processes which are again mostly unprincipled and ill understood." [Yehoshua Bar-Hillel, "Formal Logic and Natural Languages (A Symposium), *Foundations of Language V* (1969) p.15.] Is it fair to state that Bar-Hillel's challenge has, thus far, not been met?

[11]For more on the inductive-deductive dichotomy, cf. below pp. 43–44.

[12]The problems of showing invalidity are discussed in Massey [58].

[13]Black [84], Searles [85], Werkmeister [86], Hepp [87], Copi [89], Schipper and Schuh [93], Salmon [95], Carney and Scheer [97], Rescher [98], Barker [99], Freeman [101], Terrell [103], Kilgore [104], Ennis [106], Kahane [107], Michalos [108].

[14]Beardsley [88], Ruby [90, 1st edition], Little, Wilson and Moore [91], Emmet [94], Moore [102].

[16]Brody [113], Byerly [114], Annis [116], Kaminsky and Kaminsky [118], Pospesel [119], Baum [120], Blumberg [123], Ehlers [124], Manicas and Kruger [127], Simco and James [130], and Carter [131].

[17]Michalos [109], Capaldi [110] and Engel [125].

[18]Kahane [111], Thomas [115], Ehninger [117], Ruby and Yarber [90, 3rd edition], Flew [121], Geach [126], Munson [128], Scriven [129], Johnson and Blair [132], Fogelin [133], Girle *et al.* [134], Weddle [136].

[19]Purtill [112], Barry [122], and Runkle [135].

[20]It has become a practice to distinguish the artificial languages of logic from ordinary English, French, Chinese, etc., by calling the latter "natural" languages. These are languages proper, or paradigmatically. We need a term to refer analogously to arguments actually used in a first-order way to attempt to convince—and moreover used without self-consciousness about the "nature" or "structure" or some ideal of argument. The term 'natural arguments' will then distinguish such arguments from those which are invented just in order to serve as examples, and also (for the most part) from those which are self-consciously framed according to an explicit model of argument (such as arguments with numbered premises sometimes found in philosophy journal articles). It is frustrating to have to use special quotation marks to set off this term, but there is no generally accepted term with the reference we want to denote. Part of the problem is that the recognition of

the practical and theoretical significance of the difference between natural and invented arguments is just beginning to be appreciated. Woods and Walton used the term 'natural argumentation' in [91].

[21]On the last point, while most authors claim to be providing the most common and/or the most tempting or deceptive fallacies, and although lists overlap, they by no means coincide. We doubt that empirical studies have been made to discover which fallacies do occur most frequently; or even that they could be made, since there is no accepted principle of fallacy individuation.

[22]Most of the work here has simply ignored the deductive-inductive dichotomy, since it does not go nearly far enough in exposing relevant structures. More on this famous dichotomy below, cf. Section C(4).

[23]Toulmin developed this approach in [1], pp. 94-145.

[24]Scriven [140] has said, "as far as I know, there has never been an even moderately successful attempt to analyze the concept of an assumption [i.e., missing premise]" (xvi).

[25]We have subsequently learned of the work of Ennis [106] on "assumptions" (396-402). Ennis's "implicit logical assumptions" correspond to what we are calling "missing premises". (He distinguishes these from "explicit assumptions", which are undefended starting points in a line of reasoning.) He views the provision of implicit logical assumptions as a suggestive and creative activity, and offers three criteria which candidates should satisfy: gap-filling ability, credibility, and simplicity.

[26]The appearance of the *Informal Logic Newsletter*, edited by the authors, in 1978, represents a step in the direction of addressing this need. However, it is mainly an information clearing house, and prints only a few short articles. It is no replacement for a journal.

THE
INFORMAL
FALLACIES

THE NATURE AND CLASSIFICATION OF FALLACIES

Howard Kahane
University of Baltimore—Maryland County

There are two basic reasons for studying fallacies. The first is theoretical—we want to know what makes fallacious reasoning fallacious. The second is practical—we want to learn how to avoid fallacies, so as to reason more cogently. This paper reflects both of these interests.

I. THE NATURE OF FALLACY

Fallacies are errors of reasoning. We reason fallaciously when we reason invalidly, or reason from premises we should not accept, or fail to use relevant information at our disposal.

But most writers on the subject have thought otherwise. In the history of logic, fallacy has often been equated with invalidity. In fact, the predominant view has been that a fallacious argument is an invalid argument that seems to be valid.[1] But this won't do, even using the terms 'valid' and 'invalid' in a very wide sense so as to cover non-deductive arguments (as we should since the crucial steps in most everyday arguments are not deductive). Argument validity is one of three requirements of cogent reasoning, but it alone is not sufficient. We can reason validly and still be guilty of fallacy, for instance, when we reason from premises we know to be false.

Fallacious argument also has been identified with *unsound* argument, where an argument is said to be *sound* if it is valid and contains only true premises. But this won't do either. The fallacy of begging the question, for instance, is committed via sound argument whenever the begged question happens to be true.

Nor can we say that a fallacious argument is one that is either unsound or question begging.[2] For this criterion focuses too much on arguments and too little on arguers or reasoners. It is, in other words, not sufficiently pragmatic or epistemic. It is not the actual truth or falsity of premises that counts, but rather the *rationality* of believing or accepting those premises. And this is an epistemic matter.[3]

In the first place, to be rational, we don't have to be omniscient. Reasoning from a highly confirmed theory, say Newtonian mechanics, is no less cogent if later evidence discredits that theory. If false starting points automatically counted against the cogency of reasoning, almost all reasoning that is at least partly theoretical would have to be called fallacious, since most theories, no matter how well they fit what is known at a given time, eventually turn out to be false. (Of course, Newtonian mechanics, even though false, has an extremely high degree of accuracy; but that is another matter.)

And second, reasoning from true premises is not necessarily cogent, even when valid and non-question-begging. If the Chariot of the Gods theory—that long ago the Earth was visited by beings from outer space—should turn out to be true, Erich von Daniken's argument that they landed on a particular strip in South America will still be fallacious, even if they did in fact land on that strip, because von Daniken ignored readily available evidence contrary to his theory. Rational argument requires reasoning from *warranted* beliefs, not true ones.

(As an aside, it should be noted that a belief is warranted if supported by evidence and further that evidence differs from time to time and person to person. Hence, what is fallacious for one person, given his warranted beliefs, may not be for another, and what is fallacious for a given person at one time may not be at another.)

So far, it has been argued that inference validity and warranted starting points or premises are two requirements of cogent reasoning. However, there is a third requirement, which is, roughly, the use of all available or known evidence. This is the requirement appealed to, for instance, when we reject a use of Mill's Methods because relevant information has been ignored, or when we accuse someone of the fallacy of *biased statistics*, as in the famous quote from Bacon's **Novum Organum**, "Aye, but where are they painted that were drowned after their vows?" To be cogent, reasoning must therefore satisfy three criteria: it must be valid, move from warranted starting points or premises, and use all relevant information at our disposal.

(Interestingly, some philosophers, notably Carnap, claim that suppression of evidence renders inductive arguments *invalid*. I think this is a mistake, because it makes inductive validity differ from deductive validity in an unnecessary way which can be avoided simply by throwing the burden of total evidence use onto the notion of cogent rather than valid reasoning.)

II. FALLACY CLASSIFICATION

Fallacies can be divided in terms first of what makes an argument fallacious, and second what leads us to commit fallacies. The former are logical factors, the latter psychological. Let's start with the logical factors.

Since there are three requirements for cogent reasoning, there are three sorts of mistakes possible in reasoning, each one generating a fallacy class. The three thus generated might be called *invalid inference, unwarranted premise*, and *suppressed evidence*, respectively. (It should be noted that these three classes are not mutually exclusive. An argument which is invalid may contain an unwarranted premise and/or suppress evidence.)

This tripartate division is somewhat different from those usually found in logic texts. Richard Whately, in his 1826 text **Elements of Logic** comes closest to it. He divided fallacies into the *logical*—where the conclusion does not follow from the premises—and the *non-logical*, or *material*, where the conclusion does follow from the premises, but something else is wrong. He then divides non-logical fallacies into those in which a premise is 'unduly assumed' and those in which the conclusion is irrelevant to the question at issue. His category of logical fallacy corresponds to my category *invalid inference*, and his category unduly assume premise to my *unwarranted premise*. But his third category, *irrelevant conclusion*, neither corresponds to my category *suppressed evidence* nor belongs in his general category of non-logical fallacies. (It doesn't belong because arguments which validly yield an irrelevant conclusion generally fail to validly yield the conclusion at issue.) At any rate, when we divide arguments according to what makes them fallacious, we get the three broad categories *invalid inference, unwarranted premise*, and *suppressed evidence*.

The other interesting way to divide fallacies is according to the psychological mechanisms that lead us to reason poorly. Psychological mechanisms are, of course, in the realm of psychology, not philosophy or logic. The last word on them awaits a better psychology than is currently available. Still, a few things can be said on the topic; at least I'm going to try.

It seems to me that there are three sorts of psychological factors inclining us to fallacy: (1) strong emotions (needed to push us to *act* when action is necessary, but clouders of rational clarity); (2) strong desires that certain propositions be true (leading to self deception, or wishful thinking); and (3) limitations on our ability to reason cogently even when strong emotions or desires do not intrude. All of these are quite familiar, although the second, our need to deceive ourselves, is, in my opinion, the feature of human nature most overlooked by philosophers (man is *not* a rational animal, in the sense that we are not an animal programmed to reason correctly, but rather to *survive*—cogent reasoning does, of course, usually have great survival value, but sometimes not as much as does self deception or wishful thinking.)

The con artist, therefore, tries to trick us in one of three ways: by whipping up emotions, by playing on our deepest desires and fears (for instance, to be accepted by others, to live, and to be healthy), or by taking advantage of the limitations of human reasoning power. (Of course the principal con artist we have to watch out for is ourselves.) My guess as to the best psychological fallacy categories is thus that there are three such categories: *emotional confusion, conflicting desire* (or *wishful thinking*), and *rational confusion.*

Rational confusion can conveniently be divided further into three species: *complexity*, where the mind is swamped with more than it can handle; *similarity*, where lack of subtlety leads the mind to become confused by the similarity of some fallacious reasoning to the genuine article; and *false principle*, where the mind assents to a perhaps plausible but in fact erroneous principle of reasoning. (Notice again that these are not mutually exclusive. For instance, we may assent to a false principle in part because it resembles a valid one.)

III. TRADITIONAL FALLACY CATEGORIES

How does this relate to the standard or traditional fallacy categories? It seems to me that the traditional fallacy categories result from an interesting blending of logical and psychological considerations. As C. L. Hamblin says (in his book ***Fallacies***):

> A fallacious argument, as almost every account from Aristotle onward tells you, is one that *seems to be valid* but *is not* so. Two different ways of classifying fallacies immedately present themselves. First, taking for granted that we have arguments that seem to be valid, we can classify them according to what it is that makes them not so [my logical factors]; or secondly, taking for granted that they are not valid, we can classify them according to what makes them seem to be valid [my psychological factors]. Most accounts take neither of these easy courses. Aristotle's original classification tries to be both sorts at once, and there are writers even in modern times who adopt it without criticism. Of those who invent their own classifications many share this uncertainty of purpose; . . .

We thus should expect that standard fallacy classifications will contain a confused mixture of logical and psychological considerations. Take the common division into formal and informal fallacies (ignoring the fact that those fallacies usually classified as formal—*denying the antecedent, undistributed middle term*—are hardly ever committed in real life). Irving Copi uses this division in his book ***Introduction to Logic***, the most widely used logic text in history. Formal fallacies, says Copi:

... are discussed in connection with certain patterns of valid inference to which they bear superficial resemblance. ... informal fallacies [are] errors in reasoning into which we may fall either because of carelessness or inattention to our subject matter or through being misled by some ambiguity in the language used to formulate our argument.

The first of these two categories, that of formal fallacies, is a logical category—that is, arguments fall into this category because they are *invalid*, although they resemble valid arguments in form. But the second category, that of informal fallacies, is a psychological category, trading on the psychological mechanisms—*appeal to pity*, ad hominem argument, *straw man*—that trick us into accepting bad arguments. Thus, Copi's two broad categories, formal and informal fallacies, are based on different sorts of criteria—they are intended to divide all fallacies into two mutually exclusive classes, but they do not, because the rationale of one crosscuts that of the other.

This can be seen more clearly by paying close attention to Copi's definition of informal fallacy, noting that most of the fallacies he calls formal satisfy his definition of informal fallacy. He says that informal fallacies are errors in reasoning into which we may fall because [for one thing] of carelessness ...' Yet the fallacies he calls formal fallacies—*undistributed middle term, affirming the consequent*—surely are errors in reasoning, and just as surely may be committed because of carelessness. Hence, they sometimes count not only as formal fallacies, since they are invalid, but also as informal ones, since they are errors in reasoning due to carelessness.

Further, many fallacies he calls informal are just as invalid as, say, *illicit process of the major term*. Here are two examples used in several current texts to illustrate the fallacy of *equivocation*, alleged to be an informal fallacy:

Some dogs have fuzzy ears. Only man is rational.
My dog has fuzzy ears. No woman is a man.
So my dog is some dog! Hence no woman is rational.

Both of these arguments are invalid. (We can, of course, construe the second one as valid, but only by construing its first premise so as to be obviously false.) Hence, they are formal fallacies just as much as are any others. The point of labelling these arguments *equivocations* is to point out the source of the confusion which leads us to accept these fallacious arguments. But they illustrate the fact that informal fallacies that are formally invalid also are thus formal fallacies.

On my view, of course, there is nothing wrong with overlapping fallacy classes. In fact, psychological fallacy classes *must* overlap logical ones (and, as an aside, do in fact often overlap each other—a given fallacy may, for instance, trick us because it is very complicated and also because we want very much to accept its conclusion, thus being both a case of *rational confusion* and of *wishful thinking*.) One of the troubles with the division into formal and informal fallacies is that it is presented as a mutually exclusive division of fallacies when in fact it is not. (Another is that it misses the logical categories *unwarranted premise* and *suppressed evidence*.)

We also should take note here of a point briefly mentioned before, namely that it is *reasoners* who commit fallacies—arguments in themselves usually are not fallacious. Thus, there often is no one correct way to classify fallacies, even considering just logical classifications. Take the following commercial, usually said to invite the fallacy of *popularity* (a version of *appeal to authority*):

> More people in America drink Budweiser than other beer.

Taken literally, this isn't an argument, much less a fallacious one. But a strong implication of this commercial is that the listener should drink Bud. So we can restate the commercial so as to say:

> More people in America drink Budweiser than any other beer. (Premise)
> So you too should drink Budweiser. (Conclusion)

Stated this way, the argument is invalid, hence belongs in the category *invalid inference*. Yet we could just as well restate the commercial this way:

> More people in America drink Budweiser than any other beer. (Premise)
> The most popular beer is the best beer. (Premise)
> You should drink the best beer. (Premise)
> So you should drink Budweiser. (Conclusion)

Stated this way, the argument is valid, but contains a questionable premise, namely that the most popular beer is the best beer. A person who interprets the commercial this way thus cannot be guilty of *invalid reasoning*, but if he accepts the argument, he is guilty of the fallacy of *unwarranted premise*.

In the same way, most fallacies can be cast so that either they are invalid or else contain an unwarranted premise (or perhaps so that they suppress evidence).[4] How they should be classified in specific cases depends on how actual reasoners do or are likely to construe them.

Of course, in real life there often is no way to know how arguments are in fact construed, not just because we aren't privy to what goes on in other minds, but also because whatever is going on is not sufficient to decide the matter. In real life, reasoning just isn't that tight. To get someone to see that they have committed a fallacy, we thus often first have to get them to see the looseness of their reasoning, and then show that however their reasoning is tightened, it won't be cogent.

Still, certain of the standard or traditional fallacy species generally or frequently fall into particular psychological fallacy genuses, and can profitably be thought of as species of those genuses. For example, *appeal to pity* and *appeal to fear* are often best seen as resulting from *emotional confusion*; *provincialism* and *biased statistics* from *wishful thinking*; and undistributed middle and *illicit process* from *rational confusion*. Within the category of *rational confusions*, *wandering from the point* (a favorite, incidentally, of Richard Nixon, as of most politicians) usually falls within the category of *complexity*, since straying from the point usually works because the listener forgets what the original point was, and thus fails to notice the straying.

But the various instances of most standard fallacies do not tend to fall in any one broader psychological classification. *Straw man*, for instance, sometimes works because of *wishful thinking*, and sometimes because of *complexity* (the latter because the genuine position is different from the straw one in complex ways which cause confusion).

Many of the standard fallacy categories also can be profitably viewed as species of one of the three logical fallacy categories, in that many or typical instances fall under those larger categories. That is, we can construe them as often or in typical cases specifying more closely the way in which an argument is invalid, or has an unwarranted premise, or suppresses evidence. For instance, *affirming the consequent* and *undistributed middle term* can be thought of as species of *invalid inference; false dilemma* and *begging the question* as species of *suppressed evidence* (the latter because to be guilty of *straw man* one must suppress the knowledge or suspicion that the position attacked is not the genuine article).

IV. CONTEXTS OF DISCOVERY AND JUSTIFICATION

Logicians in recent times have concentrated on the notions of validity and invalidity much more than on cogent and fallacious reasoning. One reason may be that premise truth or falsity, and premise warrantability or unwarrantability, cannot be determined by the principles of logic alone. But another is adherence to the idea that philosophy is concerned only with the context of justification, and not with the context of discovery. The latter, it is often held (for example, by logical positivists), is the domain of psychologists, not philosophers or logicians.

Whatever the merits of the distinction between the contexts of discovery and justification, this much is clear. Unless we bring in background information, we cannot advance the notions of fallacious and cogent argument beyond those of valid and invalid argument. To get beyond validity and invalidity (and if we don't, what is the point of discussing fallacies?), we have to bring in the idea of known but unstated background information, first, in order to be able to assess the warrantability of argument premises, and second, to determine whether all known relevant information is stated in the premises.

In fact, background information has always been appealed to in discussing fallacious argument, although this appeal often is not explicit. It is appealed to, for instance, in the claim that an argument presents a *false dilemma*; for what makes a dilemma *false* is precisely the suppression of information suggesting a third (or fourth) alternative. (Going between the horns of a dilemma always involves the introduction of overlooked but relevant background information.) Similarly, as stated before, it is appealed to in the claim that an argument is a *straw* argument, since it is only because of relevant background information that we are justified in making such a claim.

Well, then, do we overstep the bounds of logic and philosophy when we theorize about fallacious reasoning? Not, it seems to me, when we attempt to specify what fallacious reasoning consists in, nor when we specify the logical factors which make fallacious reasoning fallacious. These are questions of methodology, and thus of logic and philosophy.

But we do overstep when we attempt to specify psychological mechanisms that lead to fallacious reasoning, and when we devise psychological categories useful in avoiding fallacious reasoning. The attempt to avoid bad reasoning requires a bringing together of philosophical and factual information, just as does every application of philosophy to real life. So the question "What makes arguments fallacious?" is philosophical; the question "What leads us to accept bad arguments?" is not.

Formal logic is just the same. A logic text which defines terms and specifies and proves the principles of valid inference remains within the realm of logic, within the context of justification. But a text which includes principles of proof strategy or aids in translating into logical notation has stepped over the line into the context of discovery. Strategy rules are applied psychological principles just as much as are psychological fallacy categories. To put the point another way, teaching people to become adept at finding and bringing relevant information to bear involves psychology, but pointing out that an argument is poor because the arguer overlooked relevant evidence, or failed to look for likely evidence,[3] involves logic, just as does pointing out that a particular argument is invalid.

One final comment. It seems to me that in order to know bad reasoning when we see it we have to know good reasoning when we see it. To know, for instance, that the *post hoc* fallacy, or the fallacy of the *small sample*, has been committed, we have to know a good deal about good inductive reasoning (for one thing, we have to know when one thing's following another is good reason for concluding it is caused by that thing). This means that learning about fallacies and teaching how to avoid them requires learning about and teaching the rules of *cogent* reasoning, including the rules of valid inference. That may not be a happy conclusion, but I think it's a cogent one.

Footnotes

[1]See C. L. Hamblin, ***Fallacies*** (London. Methuen and Co., 1970), p. 12.

[2]Alex Michalos, for instance, says roughly this in ***Improving Your Reasoning*** (Prentice-Hall, N.Y., 1970), pp. 7-10.

[3]The one important modern writer who tries to take account of epistemic or pragmatic factors is C. L. Hamblin. But earlier writers often did so also. Aristotle, for instance, thought of *begging the question* quite literally as begging the question at issue, by asking that the question itself be granted as a premise for the purpose of the argument. On his terms it makes no sense to speak of an argument as question begging in isolation of arguers arguing over that question (for then what would the *question* be?).

[4] See Richard Cole, "A Note on Informal Fallacies", *Mind*, Vol. LXXIV, N. S., 1965, pp. 432-433.

[5] Think of the person who says, "Don't tell me; I don't want to know that". Failing to look for relevant information is just as much a source of superstition as overlooking or ignoring known information. Sometimes, when a person says, "But I didn't know that", it's right to reply that he should have known it, or at least known to look for it.

PETITIO PRINCIPII AND ARGUMENT ANALYSIS

Douglas N. Walton
University of Winnipeg

The most acute problem in teaching and studying the field of informal fallacies is that lack of clear and theoretically adequate models of the fallacies makes it impossible to know or prove that what strongly seems to be a fallacy really is an argument that is incorrect, or in some sense invalid. Notoriously, it is also easy to get into unresolvable disputes about whether some evidently bad argument is an instance of one fallacy as opposed to another. But the deficiencies of what Hamblin [5] calls the Standard Treatment of the fallacies are well known. What is needed is some theory. At the same time, the unique value and appeal of the study of the fallacy domain is its potential applicability to the critical evaluation of argumentation, and therefore it is important that this theory should be strongly tied to the analysis of significant arguments.

A discouraging problem is that the quest for applicable theory might tend to take us far beyond the tidy domain of first order logic. Yet the history of the disarray that is the Standard Treatment suggests that there is little value in studying the fallacies until we achieve some general understanding of the underlying concepts of argument that are involved in the major informal fallacies. In this paper we will work towards trying to see how what is called *petitio principii* might be understood as a deficiency in arguments.

I. THE STANDARD TREATMENT OF *PETITIO*

The history of the Standard Treatment of *petitio* exhibits a pair of dualisms. First, rooted in Aristotle's own treatment, there is the tendency to see it either as an epistemological phenomenon or as a dialectical (game-theoretical) fallacy. In **Pr.An.** 64b 30 Aristotle treats *petitio* in light of his famous dictum that demonstration proceeds from what is more certain or better known: if a man tries to prove what is not self-evident by means of itself, he begs the original question (**Pr.An.** 64b 37). The account of the fallacy here is epistemic. To be the question is to violate the epistemic principle of the priority in knowledge of the premisses over the conclusion in a demonstration. In the **Topics** however, the account is set in terms of contentious disputation between two or more parties. Begging the question is said to occur where a questioner, the party who is supposed to be arguing for a certain thesis, T, asks to be granted T as a premiss to be conceded by his opponent. This second account helps to explain the apparent peculiarity to modern ears of the phrases "begging the question" and "*petitio principii.*"

The second dualism is also a common historical theme and like the first, has survived through the ages into current logic textbooks. According to the *equivalence conception*,[1] an argument is said to be circular if the conclusion is assumed as a premiss, either as an exact equivalent or in a form so close to make the two statements virtually equivalent. As Copi puts it, "... two formulations can be sufficiently different to obscure the fact that one and the same proposition occurs both as premiss and conclusion" (***Introduction to Logic***, 4th ed., New York, Macmillan, 1972, p. 83). The problem with explicating this conception is that orthographic identity is too narrow a criterion, and logical equivalence does not seem to fit either.[2] The required notion of equivalence is elusive—perhaps it could be epistemological in nature. According to the *dependency conception*, an argument is said to be circular where the conclusion is required in order to establish some premiss. That is, according to this conception of a non-circular argument, one should be able to know that each premiss is true without having to infer it from the conclusion of the argument. The problem here is to explicate the required relation of dependency. As we have stated it here, obviously the notion of dependency appears to have an epistemic flavour.[3]

The above sketch may seem to indicate that the best places to look for some adequate theory to explain this phenomenon of *petitio* are either the dialectical structures proposed by Hamblin [5], Lorenzen [12], Mackenzie [13], and Rescher [15], or epistemic logic as developed by Hintikka [8] and developed in numerous subsequent studies. However, a general tendency when confronted by something new is to attempt to reduce it to first-order logic, and the historical basis for this possibility in regard to *petitio* is found in De Morgan's ***Formal Logic*** [1].

De Morgan [1, 254] preferred the purely alethic criterion that "strictly speaking, there is no formal *petitio principii* except when the very proposition to be proved, and not a mere synonyme of it, is assumed." This perhaps is what one might expect a logician to say, but the problem is that it scarcely seems applicable to a realistic doctrine of *petitio*. De Morgan however shrewdly backed up this posture with what John Woods and I in [20] call *De Morgan's Thesis*: no syllogism begs the question. We note that a syllogism has the following two properties: (1) it always has two premisses, and (2) each premiss is logically independent of the other. Thus Woods and I suggested in [20] that De Morgan's thesis can be cast into an interesting modern form: no deductively valid argument with more than one premiss and no superfluous premiss begs the question. A premiss is *superfluous* in a valid argument if, and only if, the resulting argument is invalid if the premiss is deleted.

A main problem with De Morgan's thesis is that it appears open to some plausible counter-examples. For example, suppose I were to propose an argument to you in the form of a disjunctive syllogism [P v Q, −P, therefore Q] but clearly indicate that I mean to establish P v Q by appeal to Q through implication. Then I would have outrageously begged the question, but my argument is many-premissed and neither premiss is superfluous. De Morgan might have replied that Q is really a "premiss" of the argument,[4] but was it or not? This raises an interesting question about how we identify premisses in the evaluation of argumentation. In fact it raises a lot of interesting questions about the applicability of logic to the evaluation of arguments, for, it would seem that we might really have two "arguments" in the present instance, each "linked" to the other.

$$Q \rightarrow P \vee Q, -P \rightarrow Q$$

The issue that is also raised therefore is: how in practice do we combine arguments in chain-like sequences to form *themata*, as Geach [4] calls them, from argument *schemata*? Whether De Morgan's thesis stands or falls seems to depend on these two unresolved issues in argument analysis.

This section would not be complete without remarking on J. S. Mill's famous *dictum* that all deductive logic commits *petitio*. Mill, in his **System of Logic** [14, 120f.] argued that the first premiss of the syllogism, "All men are mortal; Plato is a man; therefore Plato is mortal" presupposes the conclusion in the sense that we cannot be sure it is true unless we are already certain that the conclusion is true. If it is doubtful that Plato is mortal, Mill argued, it is at least as doubtful whether all men are mortal. De Morgan [1, 257] threw some doubt on this argument by pointing out that it overlooked the minor premise. De Morgan, as we saw, in effect argued in flat contradiction to Mill that no valid syllogism begs the question. I suppose most of us would be inclined to avoid either of these extreme views by ruling that whether or not the syllogism above is circular depends on whether the context of argument includes (perhaps inductive) evidence for the major premiss that is independent of the conclusion, e.g. biological evidence of the mortality of animals. But this raises the perplexing question of the role of background evidence in the context of argument and consequently takes us back to the problem of what may or may not be considered a "premiss" of an argument.

II. SOME RECENT WORK

Outside of first order logic, the two most likely candidates to throw some light on the theory of circular argument would be epistemic logic, as studied by Hintikka [8] or Kripke [11], and the formal dialogues (dialectical games) of Hamblin [5]. The Hamblin game H is called a "Why-Because-Game-with-Questions" because it allows participants to advance questions of the form [Why A?] and it allows a respondent to reply with a locution of the form [Statements G, B $\supset$ A, for any B.] This game has commitment-stores for all participants, and Hamblin [5, 268ff.] imposes a dual restriction on the operation of commitment-stores that, he claims, blocks *petitio* moves: (α) [Why A?] may not be used unless A is a commitment of the hearer and not the speaker; (β) A "because" answer to [Why A?] must be in terms of statements B that are commitments of both participants (note: we henceforth restrict the number of participants to two, for ease of exposition). How does this block *petitio*? Consider a simple circle game.

	WHITE	BLACK
(1)	Why A?	Statements B, B $\supset$ A
(2)	Why B?	Statements A, A $\supset$ B

White's move at (2) is always blocked by (α) because by (β), B becomes a commitment of White by Black's response at (1). If Hamblin is right, it might seem that *petitio* can be handled, and perhaps also understood to some extent in the contexts of a game like H.

John Woods and I in [21] however have expressed some doubts about whether (α) and (β) do effectively block circle games. The game H allows for retraction of commitments by participants, and problems may occur because a participant need not be committed to all logical consequences of his commitments. Consequently, as we have shown in [21], it is possible to construct dialogue-sequences in H that may be circular, or at least where intuitions clash or are uncertain on the question of whether there is a *petitio*.[5] These problems however are not my main worry about the adequacy of rules like (α) and (β) in games like H as means of capturing the notion of begging the question. The main worry is that (β) is too strong in ruling out all justificatory moves except those utilizing premises that are commitments of one's opponent. (β) does not specifically represent the notion of begging the question at all.[6] To beg the question is to attempt to justify A either on the basis of A or on the basis of B, where B is also the basis of A as in the circle game above. Thus even if (α) and (β) do block *petitio* in H, it is highly dubious whether they jointly characterize *petitio* as a concept of argument.

Mackenzie [13] attempts to alleviate some of the problems of commitment-retraction in H in order to deal with *petitio*, but the basic problem remains that no set of rules of a game like H can adequately represent the essence of *petitio* exclusively by means of allowing or proscribing commitments of participants. *Petitio* pertains to the notion that Woods and I in [21] call groundedness, the idea that a statement A is *based on* some other statement B in the sense of being a justificatory response to [Why A?]. This notion is the heart of the dependency *petitio*.

In [21] a theory of *petitio* is offered in the framework of the Kripke semantics for intuitionistic logic [11]. In this semantics, we have a set of "evidential situations" H_i (possible states of knowledge at a given time) which are ordered as a tree. Statements A, B, C, ... take on two values, "verified" or "not-verified" at these points H_i. Essentially, [21] rules that we have a *petitio* at a pair of points H_i, H_j where H_j is a "later point" in the ordering than H_i if there is no "new information" at H_j that had not already been verified at H_i, e.g.

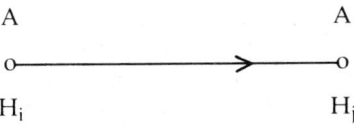

Here we would have a *petitio* in the account of [21] because the "jump" from H_i to H_j represents no "new information" at H_j. We show in [21] that according to this conception of circular inference, circles can be constructed in H even with (α) and (β).

My present thinking is that while this theory of [21] does represent a worthwhile representation of the equivalence *petitio*, it does not adequately characterize the dependency *petitio*. In order to do that we would have to express this sort of structure in the model.

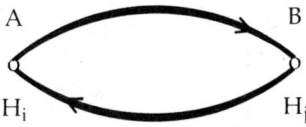

But this structure cannot be expressed in the Kripke modelling for the ordering of H_i is in the form of a tree, and it is well known that a tree is an acyclic graph. All this suggests that the Kripke model could be viewed as a special kind of graph and that *petitio* could be more generally studied in graph theory. Other factors will also suggest this possibility, and we will return to it in Section IV.

What of Hintikka's epistemic logic initially developed in [8]? The problem here is the well-known rationality assumption that a knower knows all the logical consequences of what he knows. Hamblin's H was not closed for consequences of commitments in any way and therefore had problems with commitment-retraction.[7] Hintikka's system is closed under *all* deductive consequences. One system is too weak, and the other too strong to be of much use in a realistic approach to argumentation. Indeed, it is specifically because of the strength of this rationality assumption that Hamblin [5, 238f.] rejects epistemic logic as an approach to the concept of argument appropriate to the study of the fallacies. It is not hard to see how Hamblin's remark is applicable in the case of *petitio*. As Aristotle pointed out, circular reasoning is fallacious in an epistemic context for the very reason that there is an order in the epistemic weighting of propositions. But there can be no such ordering with regard to the pair {premisses, conclusion} of a deductive argument if knowledge is closed under deductive consequence, for under that rationality assumption, both statements must be known equally by all knowers. In short, under Hintikka's assumption, there could never be anything fallacious about *petitio*.

Of course, in recent studies, e.g. [9], Hintikka has worked towards potentially more promising versions of epistemic logic by introducing the notion of a *surface tautology*. The idea is that a knower knows only the "obvious" consequences of what he knows, where obviousness is defined in terms of measures of the complexity of formulae. And then too, Hamblin [6] has tried to sort out some of the problems of assigning commitments in dialectical games by introducing the notion of an *immediate consequence* of a statement.[8] These are promising developments in needed theory.

III.
FORMAL LOGIC AND THE LOGIC OF ARGUMENT

In discussing informal fallacies it is all too easy to acquiesce in the questionable assumption that the term Formal Logic (as capitalized by Hamblin) can be bandied about as though it referred to some nicely circumscribed referent. A glance at disputes in the foundations of mathematics will reveal the quick decomposition of any such easy assumption. In the teaching of logic in philosophy departments, what often passes for Formal Logic is in fact first-order logic. A glance at heavily used textbooks like Copi's two well known primers however will indicate another fact, namely that first-order logic is often, perhaps usually, taught in philosophy departments as very much an applied logic, applied to the practice of evaluating arguments in natural language. Thus in practice, the domain of what passes for Formal Logic in many a curriculum contains much that is really of an informal nature. Let us try to be more specific. There are four basic tasks of the analysis of argumentation that the application of classical first-order logic presupposes as preliminaries.

First, classical first-order logic does not tell us whether the chunk of alleged argumentation that we are confronted with as *datum* really is an argument, as opposed to a reminder, threat, clarification, or whatever. That is, classical logic does not effect the tripartite classification *valid arguments/invalid arguments/not arguments at all*. It merely enables us to effect the former distinction. But in order to apply logic to the adjudication of an argument, we need to first decide whether we do really have an argument. This is a practical point, not merely a philosophical cavil, because a systematic sophist can always elude the force of logic by claiming, even if ingenuously, that what he advanced is not invalid because it is not an argument at all, but merely a clarification or something of the sort. Hamblin [5, 224f.] in his chapter on the concept of argument illustrates how such evasions can work in practice. So here is the first preliminary task for the application of classical logic, the identification of arguments.

Second, given a pair of statements, classical logic contains no way of determining by any exact procedure which one is the "premisses" and which the "conclusion" of an argument. Deductive implication is not symmetrical, and therefore it can make a difference if we get this backwards. If we look to the texts, they tell us that the conclusion is the statement that is "established on the basis of" the premisses, which seems to be nothing more than a blatant appeal to our intuitions in the matter. Some texts appeal to so-called *indicator* words, like "therefore." However, "therefore" is not always present, and we have no decision procedure in classical logic for telling us when an indicator word is equivalent to a "therefore".

In regard to both tasks one and two, texts often correctly point out that the premisses and conclusion must be *designated* as such. This would seem to indicate that the methodology for such designations is external to the usual methodology given for first-order logic. This move may better circumscribe the two tasks, but it does not dispose of the need for them if logic is to be applicable to argumentation.

The third task is determining what *type* of argument we are confronted with. That is, assuming that there are, in addition to deductive arguments, things like inductive arguments, or perhaps even plausible arguments[9] that are neither deductive nor inductive, how are we to identify which type a given argument belongs to? The whole issue is rife with theoretical pitfalls and unclarities, but it is more than a purely philosophical puzzler. It is a practical question of the analysis of arguments. A systematic sophist could always escape a charge of deductive invalidity if pressed, by simply claiming that his argument was merely inductive, and escape the charge of inductive invalidity in turn by claiming mere plausibility. Conversely, a sophistical deductivist could always accuse his inductively arguing opponents of deductive invalidity. If there is no objective way of really determining which type an argument is, then there is no way of effectively barring such sophistical manoeuvers and counterattacks.

A fourth task is determining how arguments are linked together in longer chains of argumentation. As we observed above in connection with De Morgan's thesis, sometimes the conclusion of one argument can also function as a premiss of a subsequent argument. Hamblin calls the resulting chain a "thread" or "development" of arguments.[10] Geach speaks of *themata* formed by chain-like linkings of individual argument *schemata*.[11] Such procedures were utilized by the medievals in applying syllogistic to extended arguments. Classical first-order logic is indifferent to this phenomenon because classical implication is transitive and no distinction is made between the cases where P implies Q mediately or immediately.

IV. ANALYSIS OF ARGUMENTS

Scriven [16] uses a method of circled numerals connected by lines to map out what one takes to be the structure of an argument in the sense of identifying the conclusions and sorting out which premises are supposed to support what conclusion, with the help of what other premises. An example [16, 78f.] will illustrate the gist of this technique.

We can be proud that (America ¹has turned the corner on the depression of the last few years.) At last (the ²many indexes of recovery are showing optimistic readings.) (The ³rate of inflation has slowed), (unemployment ⁴has more or less stabilized,) (inventories ⁵are beginning to drop,) (advance ⁶orders are starting to pick up,) and—the best news of all—(the ⁷average income figures are showing a gain.) (The ⁸doomsayers have been discomfited,) and (the ⁹free enterprise system once more vindicated.)

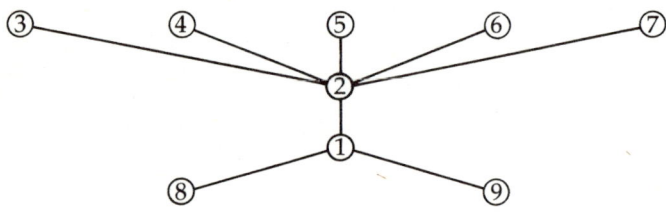

According to the reading given by Scriven, ① is the key conclusion. ③ , ④ , ⑤ , ⑥ , and ⑦ are marshalled in support of ① , and summed up in ② . ⑧ and ⑨ are two further assertions drawn from ① , but not from each other.

Where "each of the considerations has weight only in conjunction with others" (p. 80), Scriven ties the premises together with a horizontal bracket (p. 42), e.g.   ① + ② + ③ − ④   (the + reads "with" the −

"despite"). Geach [4] uses a similar method. If we have a sound argument from A and B to C, and another from C and D to E, and yet another from E to F, then according to Geach, we have a sound argument from A, B and D to F. Geach [4, 66] uses this form of diagram.

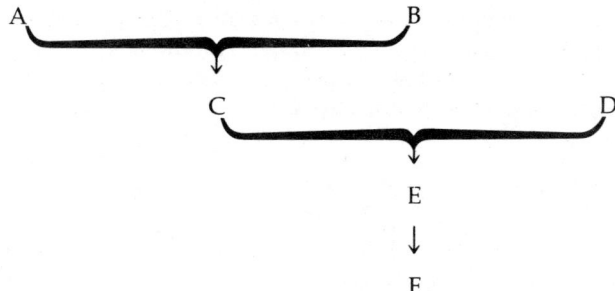

Ehninger [3] uses a similar technique of arrows to join argument chains and clusters.

This general type of method is extremely useful in the practice of evaluating arguments, and helps to provide a working basis for helping with our four tasks of applying logic in the analysis of arguments.[12] It is particularly helpful in enabling us to sort out which are the premises and conclusions of an extended argument, and in linking arguments up in a chain-like fashion (the fourth task) as indicated in the illustration from Geach given above. However, the use of this method raises many theoretical questions about what we are doing in using it, and I would like to raise some of these questions.

The Scriven-Geach-Ehninger technique can easily be adapted to express cyclic patterns of argument, e.g.

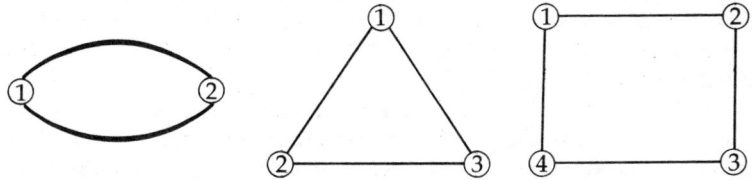

But how are we to understand such cycles? Do they represent the fallacy of *petitio*, or might they sometimes be non-fallacious in nature? Could they possibly be ways of modelling instances of the dependency *petitio?* It seems to depend on what is meant by the points (circled numerals) and steps (lines).

V. DIGRAPHS OF ARGUMENTS

The various diagrams above are obviously reminiscent of the familiar diagrams of graphs in graph theory—see Harary [7]. The question then is whether an argument analysis of the Scriven-Geach-Ehninger type could be viewed as a directed graph. A *directed graph* or *digraph* consists of a finite nonempty set V of *points* (nodes, vertices, etc.) and a set X of ordered pairs of points. These ordered pairs are called *arcs* (directed lines, edges). Digraphs are usually drawn as points connected by arrows, as in the illustration below representing the digraphs with three points and three arcs.

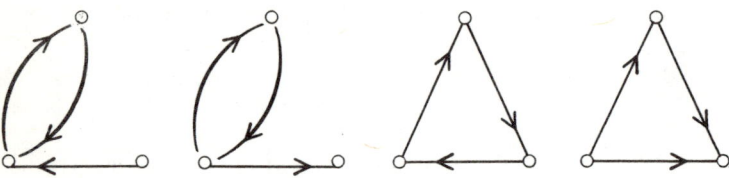

A *loop* is a line that joins a point to itself, e.g. ⟲ . If more than one line joins two points, e.g. ⇄ , it is called *multiple lines*.

An arc of a digraph can be thought of as a binary relation, and as such its properties are relatively weak as relations go. It need not be reflexive—that is, we can have loops or not as we wish. It need not be transitive—just because there is a line from v_i to v_j and one from v_j to v_k, there need not be a line from v_i to v_k. And it is not symmetrical or asymmetrical. In a (nondirected) graph it need not even be non-symmetrical but in a digraph, it is non-symmetrical—that is, if there is a line from v_i to v_j, there may or may not be a line from v_j to v_i.

The relative non-committal nature of digraph theory accords well with what I take to be the spirit of the Scriven-Geach-Ehninger method in not requiring symmetry or asymmetry. Let us now turn to the key questions of transitivity and loops.

A *walk* of a graph G is an alternating sequence of points $v_i \in V$ and lines $x_i \in X$, $v_0, x_1, \ldots, v_{n-1}, x_n, v_n$, beginning and ending with points, and where each line is incident with the two points immediately preceding and following it (See [7, 13]). In terms of the theory of relations, the notion of a walk permits a kind of transitive closure[14]—if there is a line from v_i to v_j and a line from v_j to v_k, it does not follow that there is a line (arc) from v_i to v_k, but it does follow that there is a walk from v_i to v_k. In terms of argument analysis, what this means is that if there is an argument from one premiss-set to a conclusion, and then from that conclusion (as premiss) to a second conclusion, and so forth from that point to some end statement, then we can say after Hamblin that there is a "thread" or "development" of arguments, as we put it earlier a "chain" of argument, from the initial premisses to the end conclusion. In other words, digraph theory models the structure of our fourth task of argument analysis very nicely. In graph theory we have the distinction between a walk which may have many intervening arcs, and a "single-step" arc v_i, v_j where there are no points v_k between v_i and v_j. In the theory of argument analysis, we are back to the distinction—tacitly relied on by De Morgan—between a single-step argument and chain-like collocation of argument steps to produce a complex argument.

Suppose we give in to the temptation to think of an argument in this graph-theoretic way. Each argument is composed of a set of points which represent bundles of statements that are "premises" or "conclusions." A pair of points is joined by a directed line that represents the step from the initial point (the premisses) to the end point (the conclusions). A walk, or sequence of arcs, represents a thread of argumentation, a sequence of premisses and conclusions joined together to form a longer chain of reasoning. Now some pregnant philosophical questions: what do the points and the arcs really represent in arguments? By focussing attention on the *petitio* these questions may be sharpened.

In graph theory, a walk is said to be *closed* if $v_0 > v_n$, and *open* otherwise. A closed walk with n ≥ 3 distinct points is called a *cycle*.[15] For our purposes it is nice to define a *circle* as a walk that is a loop, multiple lines, or a cycle. In the context of argument analysis, a natural doctrine of circular argument may be formulated as follows. A loop represents an equivalence *petitio* and a circle of n ≥ 2 represents a dependency *petitio*. In looping, one has argued that P on the basis of P. In cycling where n ≥ 2, one has started a chain of argument with initial premiss P and "arrived back" at final conclusion P. Interestingly, all circles come out to be "purely formal" circles of this sort, just as De Morgan would have liked, but how "strictly logical" a theory of this sort really is depends on how we interpret the points and lines. It would not do, for example, to interpret an arc as a logical implication because implication is transitive. Of course, as far as the theory of directed graphs is concerned, we have virtually limitless freedom in how we wish to interpret points and lines. But the question is rather one of finding practical interpretations that are well suited to the analysis of arguments.

VI. WHAT'S WRONG WITH ARGUING IN A CIRCLE?

Eberle [2] proposes that we make epistemic logic more realistic by adopting as a binary operator on statements a relation I (P, Q) to be thought of as representing an actual inference carried out by some inferrer. He then proposes a system, proved sound and complete, that has the following property in place of Hintikka's rationality assumption: if P is known to be true and Q is inferred from P, then Q is also known to be true. This system is the one I would find most applicable in the analysis of actual argumentation because in an argument analysis we are concerned with attempting to determine which inferences an arguer has actually made. The main problem with Eberle's system for this application is that I is transitive: if I (P,Q) and I (Q,R) then I (P,R). But it seems to me that transitivity is a highly significant sort of rationality assumption in itself. Why should all actual arguers be unerringly transitive in their inferences, any more than they are in their preferences?

I propose instead that we define $I_D(P,Q)$ initially as a non-transitive binary relation, the relation of *direct inference* of Q from P, and then define its transitive closure $I_I(P,Q)$ as follows: $I_I(P,Q)$ if, and only if, there are statements $R_1, R_2, \ldots, R_i$ such that $I_D(P,R_1)$ and $I_D(R_1,R_2)$ and ... and $I_D(R_i,Q)$. We read $I_I(P,Q)$ as the relation of *indirect inference* of Q from P. Now we can think of a graph yielded by an argument analysis as a specimen of an epistemic inference. A "single-step" arc from v_i to v_j is a direct inference from the statement (premiss) v_i to the statement (conclusion) v_j. A "multi-step" walk from v_i to v_j represents an indirect inference from an initial statement (initial premiss) mediated through a sequence of argument steps to an end statement (end conclusion).

Putting inference in this epistemic framework helps us to see what can be wrong or fallacious about cyclic patterns of inference. If we take the basic notion of inference I_D as above, along with its ancestral relation I_I, there is nothing demonstrably wrong with circular inferences as actual inferences carried out by some inferrer. What can be wrong, as the philosopher told us,[16] is that if there is an epistemic order of propositions, circular reasoning may violate this ordering. What is the right ordering? This question is perhaps the most fundamental one for epistemology, and I will not by any means try to answer it here, but only to point out that there are possible orderings that exclude circles. For example, a linear ordering of the points v_i excludes circles. A tree-like ordering of the points v_i excludes circles. And so on. The idea is that we can have different species of *rational inference* as we add different conditions on the basic epistemic digraph which by itself only represents the actual inferences of a reasoner, bereft of various levels of rationality that may be epistemically imposed on his sequence of inferences.

In other words, we are in the pleasant position of being able to concur with Aristotle that *petitio* can be seen as an epistemic fallacy. The lines of an argument digraph can represent the epistemic dependency of the end point on the initial point, and therefore in any sequence of inference through a chain of propositions which represent linkings of premisses and conclusions there can be an ordering of the propositions such that each is better known than its successor. Let us say for example that we have a linear ordering of propositions $v_0, v_1, \ldots, v_k$ such that each $v_i < v_j$ is less well known than v_j. What this implies is that the chain of inferences at any arbitrary point v_i can never form an inference with the pair $<v_i, v_j>$ if $v_j \leq v_i$. Thus circles of either the dependency or equivalence variety can be ruled out by setting epistemic priorities.

It is worth emphasizing that, so construed, circular inference is not identical with the fault of choosing premises that fail to be more certain than one's conclusion. Rather, circular inference is one particular species of this more general shortcoming of arguments. It is that species where either Q is inferred from itself, or from some P that is also inferred from Q.

Footnotes

*Research for this paper was supported by a grant from the Social Sciences and Humanities Research Council of Canada.

[1] This phrase is taken from [19].

[2] For an elaboration of these points, see [19].

[3] A detailed account of these two conceptions is given in [19].

[4] For more on this issue, see [20].

[5] The reader should look to [21] for a detailed account of these dialogue-sequences.

[6] (β) represents the inadequacy of the arguer's premisses to convince his opponent by argument. In part VI we will distinguish between the analogous shortcoming in an epistemic context (as opposed to the dialectical context here) and the fallacy of *petitio principii*.

[7] See [21].

[8] See also Mackenzie [13].

[9] See Nicholas Rescher, **Plausible Reasoning**, Assen/Amsterdam, Van Gorcum, 1976.

[10] [5, 229].

[11] [4, 66].

[12] Extended examples of application of this method to the analysis of argumentation are given in Vignaux [17, 294] and Johnson and Blair [10, 177f.].

[13] A useful text on digraph theory is Frank Harary, Robert Z. Norman and Dorwin Cartwright, **Structural Models: An Introduction to the Theory of Directed Graphs**, New York: Wiley, 1965.

[14] Transitive closure of inferences is defined in part VI.

[15] See [7, 13].

[16] In part I.

References

[1] Augustus De Morgan, **Formal Logic**, London: Taylor and Walton, 1847.
[2] Rolf Eberle, "A Logic of Believing, Knowing, and Inferring," **Synthese**, 26, 1974, 356–382.
[3] Douglas Ehninger, **Influence, Belief, and Argument**, Glenview, Illinois: Scott, Foresman and Co., 1974.
[4] P. T. Geach, **Reason and Argument**, Oxford: Blackwell, 1976.
[5] C. L. Hamblin, **Fallacies**, London: Methuen, 1970.

[6] _____, "Mathematical Models of Dialogue," *Theoria*, 37, 1971, 130–155.
[7] Frank Harary, *Graph Theory*, Reading, Mass.: Addison-Wesley, 1969.
[8] Jaakko Hintikka, *Knowledge and Belief*, Ithaca, N.Y.: Cornell University Press, 1962.
[9] _____, "Surface Information and Depth Information," in *Information and Inference*, ed. Jaakko Hintikka and Patrick Suppes, Dordrecht: Reidel, 1970, 263–297.
[10] R. H. Johnson and J. A. Blair, *Logical Self-Defense*, Toronto: McGraw-Hill Ryerson, 1977.
[11] Saul Kripke, "Semantical Analysis of Intuitionistic Logic I," in *Formal Systems and Recursive Functions*, ed. J. N. Crossley and M. A. E. Dummett, Amsterdam: North-Holland, 1965, 92–130.
[12] Paul Lorenzen, "Ein Dialogisches Konstructivitätskriterium," in *Infinitistic Methods*, London: Pergamon Press, 1961, 193–200.
[13] J. D. Mackenzie, "Question-Begging in Non-Cumulative Systems," *Journal of Philosophical Logic*, to appear.
[14] J. S. Mill, *A System of Logic*, London: Longmans, Green, 1843.
[15] Nicholas Rescher, *Dialectics*, Albany: State University of New York Press, 1977.
[16] Michael Scriven, *Reasoning*, New York: McGraw-Hill, 1976.
[17] Georges Vignaux, *L'Argumentation*, Genève: Librairie Droz, 1976.
[18] Douglas Walton, "Mill and De Morgan on Whether the Syllogism is a Petitio," *International Logic Review*, 8, 1977, 57–68.
[19] John Woods and Douglas Walton, *"Petitio Principii," Synthese*, 31, 1975, 107–127.
[20] _____, *"Petitio* and Relevant Many-Premissed Arguments," *Logique et Analyse*, 77–78, 1977, 97–110.
[21] _____, "Arresting Circles in Formal Dialogues," *Journal of Philosophical Logic*, 7, 1978, 73–90.

FORMALISM: PRO AND CON

WHAT IS INFORMAL LOGIC?

John Woods
The University of Lethbridge

Theories benefit from formal methods in well-known ways. Such methods can yield up perspicuous structural representations; they can furnish taxonomic and definitional depth and clarity; they sometimes can provide for the effective or anyhow non-constructively successful recognition of a theory's target-properties. Perhaps it may be assumed that a body of knowledge is non-trivially eligible for formal treatment when (1) the objects of theory enter into interesting systematic interconnections expressible in functional or quasi-functional ways; and (2) such interconnections obtain or not, as the case may be, under semantic suppression of the connected items. Functional connectedness amidst semantic suppression may, then, reasonably be said to be the essential basis of the fruitful deployment of formal methods. But there may be other features of a theory's domain that bid welcome to formal techniques. In logic, for example, we deal with infinite domains of sentences (or, if you like, propositions or thoughts) and with unbounded sentential and argumental operations upon them. The law-like interconnections that logicians seek to record in the appropriate metatheorems are general in their sweep; and generality over infinite subject matters is known to be clearly and manageably expressible in formal terms. Indeed, to such *philosophical* questions as, "How, with only finite attention and finite resources, can such infinite generalities be wrought or thought?", the formal logician can sometimes proffer the notion of finite axiomatizability within complete, consistent and sound logistic systems.

Suppose now that someone identifies "The cow kicked the mule, hence the cow kicked the mule" as a circular inference. He might go on to add, quite correctly, that so too is any inference similar in salient respects. While perfectly true, this is an oblique generalization; its infinite upshot is only alluded to, not identified. The formal logician has a better way of saying much the same thing: "Any inference [A, hence A] is circular." It is better because the use of variables accommodates infinite upshot with finite resources, but also because, for any object you like, it is infallibly and mechanically ascertainable in at most finite time whether or not it is an inference, [A, hence A.] So we have recursive enumerability and decidability.

A further advantage of this sort of formal approach is that it demonstrates that, and the extent to which, possession or lack of target-properties (e.g. circularity) is not a matter of parochial semantic status, and not a matter either of parochial contextual and pragmatic features. Now, to be sure, beyond the frontiers of simple first-order theories, target-properties can be expected to involve semantic, pragmatic and contextual considerations that first-order theories need not and cannot represent. But the fact remains, if such non-first-order theories are to be formal theories, there will always be parochial semantic, pragmatic and contextual irrelevancies, and they will always be suppressed.

When you have formal representability, recursive enumerability and decidability over infinite domains of semantically suppressed primitive items systematically dispersed in functional ways, then you have, or are close to having an abstractly mathematical system, M. The system M may *itself* be a good candidate to be an object of study, an object of foundational or metamathematical scrutiny, itself organized formally and axiomatically. If the theory that M actually expresses is a logical theory—for example, a theory of argument-circularity—it may also be true that, considered as an object of foundational attention, M is metamathematically interesting in ways that have little to do with the fallacies. It is useful to keep clearly in mind that the original motivation and the metamathematical motivation may be quite different, and that they can involve M in essentially different ways, and that some of what your metamathematical investigations may disclose about M will do nothing to illuminate the *petitio*. Of other such lessons, however, I am not so sure. Would it, for example, be helpful for a clear understanding of the *petitio* that M be metamathematically determined to be isomorphic with Kripke's intuitionistic semantics or with Hintikka's new system of dialogic?[1] Quite possibly so. In any event, should M be a fit object of metamathematical attention, two kinds of illumination may be forthcoming in the metamathematics. (1) There may be a richer appreciation of M's *subject matter*, and (2) one may learn something of M's own expressive and demonstrative limitations and capacities.

Plainly, then, it would seem that being a mathematical system is not necessarily a liability for a theory of the fallacies, but, on the other hand, it certainly is not necessary that fallacy-theory be a mathematical system if the fallacy-theory is to be instructive or even deeply correct.

One of the most persistent prejudices against the use of formal methods in the treatment of the fallacies comes from confusing the use of formal methods as such with *formalization*, that is, with the construction of what Church calls logistic systems. Now it is quite true that a clear understanding of a logistic system is no guarantee of a clear understanding of the body of doctrine that it formalizes. At times a formal definition will be little more than "a platitude restated in pedantic obscurity". (Consider, for example, the usually formal definition of the continuity of a curve.)[2] Then, too, there are important limitations in relation to motivating intuitive concepts. We cannot axiomatize without residue either the intuitive notion of positive integer or the basic notion of set.[3] It would surprise me greatly if consistent, complete and sound logistic systems were found to be intuitively adequate to such notions as *argument, part-whole, expertise* and *burden of proof*, all of which, after all, are needed for fallacy-theory. But what are we to take from this? I believe that we may say that if formal treatment involved nothing but the construction of axiomatic logistic systems, then the theory of the fallacies almost certainly could not in any non-trivial sense be a formal theory. We know, however, that axiomatic formalization does not exhaust formal treatment; and so the prejudice against the formal pursuit of the fallacies requires a different justification.

It is, of course, quite obvious that some bodies of doctrine do not formalize deeply, extensively or at all. This may indicate that certain objects and their target-properties are by nature insusceptible of all but trivial formal management. It may *also* indicate merely that extensive use of formal methods is premature, that the intuitive terrain is taxonomically too unsettled and definitionally too unboundaried, and its basic conceptual geography too little known to admit of formal reorganization. I have no hesitation in saying that much of fallacy-theory fails of formal treatment because the data to be formalized are not intuitively well-enough understood, are taxonomically and definitionally unfocused, and so on. But I cannot bring myself to a more austere skepticism than this. Of the dozen or so fallacies that Douglas Walton and I have been studying recently[4] there is not one case in which the investigation did not benefit from the application of formal methods. Graph theory and intuitionistic logic are, we think, helpful in modeling circularity;[5] causal logic fixes perspectives for the *post hoc*;[6] Hintikka's system of dialogic gives an interesting representation of dialectical exchange; Routley's consistent and complete system of dialectic illuminates certain features of the *ad hominem*;[7] various constructions of eretetic logic work well for Many Questions;[8] and so on. In our own work, Walton and I have been impressed to discover two particular advantages in the deployment of formal resources. One is the provision of clarity and power of representation and definition. The other is provision of verification *milieux* for contested claims about various fallacies. As for representational or definitional headway, we repeat that circularity models well in Kripke's intuitionistic semantics, and that a reasonable notion of evidential cumulativeness is also there definable.[9] Then, too, Burge's formal theory of aggregates furnishes one with a quite powerful (though not effective) command of part-whole relations, and the theory of composition and division plainly benefits from this.[10] (By way of illustration, I develop this last point in the Appendix.) Berger's connectibility logic provides a notion of causal connectibility defined over a four-dimensional differentiable manifold. It has occurred to Walton and me that this definition allows one to represent the *post hoc* as a modal fallacy of arguing from possibility to actuality.[11]

I said that formal systems may also serve as verification (and for that matter, falsification) *milieux* for disputed claims about the fallacies. For example, if one places inference-structures (as opposed to entailment-structures) in a classical logic, then Mill's contention that the syllogism is a circular form of inference seems next-to-confirmed.[12] But if your base logic is relevant logic, it is next-to-disconfirmed. The usefulness of this observation may consist in its suggestions that relevant logic is a better (perhaps only marginally better) logic of inference than classical logic, and perhaps, too, that classical logic is a better logic of entailment than relevant logic. Or, to take another example, if you reconstruct the part-whole relation mereologically, then no inference from part to whole is a fallacy, and a great many inferences from whole to part which are not fallacious would be classified as fallacies nevertheless.[13] From which it may be concluded that the salient part-whole relation is not mereological (any more than it is set-theoretic).

In addition to the formal devices that I have already spoken of, Walton and I have found it necessary variously to repose the theoretical burdens of the fallacies in probability theory, acceptance theory, epistemic and doxastic logic, and rationality theory.[14] Few of these systems are fully axiomatic; fewer still offer much promise of recursive enumerability and decidability with respect to basic intuitive notions or target-properties. Each is, in its own right, an object of some degree or other of controversy and uncertainty. None is a perfect instrument of analysis. At this stage of their theoretical development, the fallacies do not admit of and do not need grand axiomatic reorganization; for one thing, you cannot reorganize what is not yet organized. But this no more counsels against formal techniques for the fallacies than the open-endedness and non-effectiveness of aggregate theory counsels against Burge's impressive formal treatment of that subject.

Speaking of aggregate theory, it is useful to point out that it is an empirical theory, not a logical one[15] and it is probably true to say that rationality theory, of the sort that Kyburg, Harper, van Fraassen and others are working on, will devleop more as a contribution to epistemology than to logic. Then, too, it hardly seems correct to associate dialogical systems such as those of Hamblin and Hintikka less with games theory than with logic. *This leads me to suggest not that the mature theory of the fallacies is a branch of logic that is essentially informal, but rather that the mature story of the fallacies is a branch of formal theory that is essentially extralogical in major respects.* The formal theory of the fallacies is not (just) logic. What else it is or may prove itself to be remains to be seen; but it is vastly unlikely that it will not involve those extralogical theories that we have just met with.

More than one of my colleagues has met this suggestion with a certain civilized horror. Is not rationality theory at least in part a psychological venture? Indeed it is, and the author has (brazenly or inadvertently, who's to know) saddled himself with a gross and ridiculous Psychologism. Did the great Frege rail against this heresy to no avail?

Psychologism. Let me identify a rather silly kind of blunder. One commits it by reasoning as follows: If T is a theory with domain D and if $\alpha_1, \ldots, \alpha_n \in D$ and are (or are representations of) mental contents, and if the consequences of T include sentences in the form: $[A(\ldots\alpha_2\ldots)]$, then such are psychological sentences and T is given over to Psychologism. Perhaps charity would recommend that this appalling argument not actually be attributed to anyone, but accuracy commands that mention be made of Carnap's criticism of Russell in "Empiricism, Semantics and Ontology". It will be recalled that Carnap was complaining of Russell's "Psychologistic" view that propositions are mental events.

Historically speaking, a theory was Psychologistic when three conditions were met:

(i) The truth of the consequences of T was subjectively determinable (or some such thing).

(ii) This was so because such propositions were implicitly about one's own states of consciousness to which one alone had reliable access.

(iii) The consequences of T included ascriptions of certain target-properties that were *normative* (or some such thing); for example, validity, correctness or consistency.

Is this, then, what an interest in rationality theory commits me to? I don't see that it is; I doubt that it is; but I am prepared to wait and see.

I have been saying something about why I think that the fallacies are usefully pursued within formal theory. On the evidence, the fallacies involve systematic interconnections unvarying among items that themselves admit of some degree of parochial semantic, pragmatic or contextual variation. Thus the domain of theory can reasonably be taken to be an infinitude of abstract (and somewhat idealized) objects interacting in somewhat functional ways. Such a set-up *suits* even if it does not call out for formal address.

Of course it is true that all our actual arguments are contextually unique, and it is possible, I suppose, that their formal construal will leave un- or under-represented those properties that make for such uniqueness, and moreover that these are the crucial properties for determining the adequacy or otherwise of those actual, real-life arguments. But if contextual or semantic peculiarities always carried the evaluative day, one could bid farewell to fallacy-theory, never mind whether or not formally transacted. And if contextual or semantic peculiarities always carried the day, then farewell, too, to first order logic. Unless and until it can be shown that an argument's status regarding e.g. circularity is inextricably bound up with its individual peculiarities, whereas its status regarding e.g. deductive validity escapes such idiosyncratic particularities, then the undoubted contextual singularity of real-life arguments no more dismisses the one preoccupation than the other from serious theoretical prospect. (No more than would the undeniable contextual peculiarities of the dome of St. Peter's basilica prohibit the attention of theoretical metallurgy while welcoming the attention of theoretical physics.)

The contextual peculiarities of real-life arguments make for a point of some importance nevertheless. The intuitive notions of argument, inference and the like are not (so far as I can believe) recursively enumerable, to say nothing of recursive. And it is arguable that the connectives of truth-functional logic are, each of them, less than wholly adequate reconstructions of their counterpart conjunctions in English. Thus we have it that the objects of truth-functional logic are not, strictly speaking, real-life arguments made up of real-life statements, involving the operation of real-life conjunctions. No, the objects of truth-functional logic are formal objects which, in one or other allowable way, represent the real thing. It is always nice to question as to the success or failure of such a formal theory's fit with reality, that is to say, as to the success or failure of its *application*. The same is true of any formal theory of the fallacies; it may always be open to question whether any such theory has a wholly satisfying application in the realm of actual argumentation. But that no more disqualifies the formal theory of the fallacies than it does truth-functional logic.

It is worth repeating that almost certainly axiomatization is not yet (if ever it would be) the way for fallacy-theory to develop.[16] This seems to me not to be the time for logistic systems. But neither is it sound to commit the story of the fallacies to what we might call "situational logic", that is, to an enterprise of merely anecdotal casuistry which over-concentrates upon the non-recurring case in all its ineffable singularity. What the fallacies now need is not the radical Fox but the moderate and common-sensical Hedge-hog.

I have been assuming throughout that the principal content of what is so often called "informal logic" is the fallacies—*accentus*, affirmation of the consequent, ambiguity, amphiboly, *argumentum ad baculum, argumentum ad hominem, argumentum ad ignorantiam, argumentum ad misericordiam, argumentum ad populum, argumentum ad verecundiam*, begging the question, composition, denial of the antecedent, division, equivocation, *ignoratio elenchi*, illicit process, many questions, *non causa pro causa, non sequitur, post hoc, ergo propter hoc, quaternio terminorum, secundum quid*,[17] as well as the several fallacies that Alex Michalos has invented and the few that he commits; and of course a theory of argument that is sensitive to all this complexity. If this has been a tolerable assumption, then I have an answer to the question with which we began, "What is Informal Logic?" *Nothing is*. The theory of the fallacies is not logic, though it includes some logic, indeed quite a bit of logic; and the theory of the fallacies is not only at its best as a formal theory, it is difficult to see how the suppression of its formal character could leave a residue fully deserving of the name of theory.

Now, this is not to deny that, on a quite different interpretation of "informal," there do exist perfectly legitimate and familiar instances of informal "logic". An analogy with mathematics might serve the point at hand. Mathematics that is done in the usual, workaday way, that is to say, in ordinary mathematical English and prior to any axiomatic treatment, is said to be informal mathematics. There is no reason to deny to fallacy-theory this same kind of informality. In both kinds of case, informality is a pre-axiomatic affair, and I have been at some pains to persuade the reader that the construction of logistic systems is not by any means the only, or best, way to employ formal methods.

And, of course, further semantic confusions still are possible. One might take a fallacy-theory to be informal just because it is not worked up within a strictly *deductive* framework—or, for that matter, a framework of *inductive* logic. (Nor should it be.) Or, one might suppose a treatment of the fallacies to be informal if it stresses the complexities of the *application* of theory to the on-going scene. Further still, the fallacies might commend themselves to our attention from the point of view of *praxis*—as manuals of self-help for the ratiocinatively insecure.

But note. *These* enterprises do not preclude the quite vigorous exercise of what I have been calling formal methods. On the contrary, they very much *require* such an exercise if they are to attain the generality or power that commands serious philosophical attention.

Appendix

Illustration of the use of Formal Theories in the Development of a Concept of the Part-Whole Relation Adequate to an Analysis of the Fallacy of Composition and Division.

The fallacies of Composition and Division involve misinferences from parts to wholes and from wholes to parts. An account of such fallacies must contain an account of relations of part to whole.

1. *Are Wholes and Their Parts the Same as Sets and Their Elements?*

The short and easy answer is No. Sets are not spatio-temporal objects and do not admit of being acted upon or perceived. Sets are abstract; and where their members are concrete, most interesting composition and division questions are answered automatically (and, too often, erroneously) in the negative. Sets and their members do not simulate a type of whole-part set-up adequate to the analysis of composition and division.

2. *Are Wholes and Parts the Same as Suppesian Bodies and their Parts?*

A better choice than set theory for the reconstruction of the part-whole relation would be something like mereology in the fashion of Lésniewski.[18] For one thing, a significant aspect of the motivation of mereology was to hit upon a theory of classes, not only as actual collections of their objects, but also as collectivities which would mirror some of the features of collective as opposed to distributive predication. An interesting development of what is basically a mereological theory can be found in Noll's contributions to a theory of bodies adequate for the foundation of mechanics, and in an extended and refined version of Noll's account, due to Suppes, a version which we shall follow here.[19] One needs a workable notion of part in order to plumb the "logical" structure of composition and division. Bodies, paradigmatically perhaps, have parts, and so one may expect a theory of bodies to say something useful about parts.

In the Noll-Suppes theory 'π' can be taken to designate the part relation. The following definitions ensue:

If B π A and C π A then A is an *envelope* of the pair-set {B, C}. A is the *least envelope* of {B, C} if, and only if, A is an envelope of {B, C} and for any X that is an envelope of {B, C}, A π X.

A is a *common part* of {B, C} if, and only if, A π B and A π C. If A and B are bodies, they are *separate*, if , and only if, they lack a common part. If A and B are bodies, then A is a least part of B if, and only if, A π B and there is no body C such that C π A and C = A. Body A is the *greatest common part* of {B, C} if, and only if, A is a common part of {B, C} and for every body X, if X is a common part of {B, C}, X π A.

Furthermore, if A is the *least envelope* of {B, C}, then the set-theoretic union of B and C is identical to A; A, then, is the *join* of B and C. If A is the greatest common part of {B, C}, then the set-theoretic intersection of B and C is identical to A; A, then, is the *meet* of B and C. (The operations of joint and meet are partial.)

If $A_1, \ldots, A_n$ are parts of B_1, if $A_1 U \ldots U A_n$ exists, and if $A_1 U \ldots U A_n = B$, then $\{A_1, \ldots, A_n\}$ is a *finite dissection* of B.

Finally, a *binary structure* W = <W, >π is a *structure of bodies* if, and only if, the following axioms obtain for all A, B, C and D in W:

1. A π A. (Reflexivity)
2. If A π B and B π A, than A = B. (Symmetry)
3. If A π B and B π C, then A π C. (Transitivity)
   (Note, then, that the part-relation gives a partial ordering of the bodies in W.)
4. If A and B have a common part, then they have a greatest common part.
5. If A and B have an envelope, then they have a least envelope.
6. If A is part of B and A $\neq$ B, then there is a body C in W such that B is the least envelope of {A, C}.
7. Every body has a least part.
8. Every body has a finite dissection of least parts.
   (Note, too, that axioms 7 and 8 commit the theory of bodies to atomism; by 7 every body contains at least one atom, and by 8 every body is composed of finitely many atoms.)

The Noll-Suppes account of part is more adequate for the present purpose than the standard idea of set-membership. But it is still somewhat too heavy-handed. I shall mention just two examples. (1) The axioms for the part-relation provide that properties true of all parts *always* compose. But what makes composition so interesting is that it is *sometimes* a fallacy, the fallacy precisely of supposing that what is true of all parts is true of the whole! The Noll-Suppes notion of part entails that composition is not a fallacy; and that, indeed, is too awkward a consequence. (2) Moreover, in the atomism of this theory there are additional difficulties. Intuitively, if a chain is pure gold so are its parts. But it is not required of *Suppesian* parts that the smallest parts of a chain be its links. In particular, the Noll-Suppes theory allows the *chemical* atoms of the links to be parts of the chain. As a consequence, a startling number of attributes which we could confidently expect to compose and divide fail to do so; and fallacies are ascribed where none there are.

3. *Are Wholes and Parts the Same as Burgean Aggregates and their Components?*
The notion of an aggregate and its components has been recently developed by Tyler Burge.[20] For current purposes, aggregates can be linked to first-order sets, i.e. sets containing only individuals as members. However aggregates are importantly different from sets. For example, the empty set is *not* an aggregate and no singleton is an aggregate; from the point of view of aggregate-theory, a singleton of which the sole member is an individual just is that individual. As will shortly become clear, aggregates also exhibit sharp differences from mereological classes and Suppesian bodies. We turn now to the formal development.

In place of the set-theoretic notion of membership, given by the ϵ relation, aggregate theory speaks of *componentiation*, i.e. being a component-member, symbolized by 'α.' Wherever x bears to y the α-relation, we could say, somewhat barbarically, that x *componentiates* y. The predicate expression, 'ax,' for 'a is an aggregate,' can be defined thus:

(i) $ax \longleftrightarrow (\exists z)(\exists y)(z \alpha x \,.\, y \alpha x \,.\, z \neq y)$

That is, x is an aggregate if, and only if, at least two objects are component members of it. A definition is also given for aggregate-abstraction:

(ii) $\hat{x}/\emptyset x =_{df} (iy)(x)(x \alpha y \longleftrightarrow (\exists z)(\emptyset z \,.\, x \alpha z))$

The counterpart of set-comprehension is:

(iii) $(y \alpha \hat{x}/\emptyset x) \longleftrightarrow (\exists z)(\emptyset z \,.\, y \alpha z)$

An explicit denial of the empty aggregate can be got from:

(iv) $-(\exists y)(y = \hat{x}/x \neq x)$

Analogous to the principle of extensionality for sets we have:

(v) $(az \,.\, ay) \to (y = z \longleftrightarrow (x)(x \alpha y \longleftrightarrow x \alpha z))$

where the antecedent reflects the intention of the theory that not everything be an aggregate. In fact, our next principle makes the point more precisely: only individuals are components of aggregates.

(vi) $x \alpha y \to Ix$,

where individuals may be taken to be their own components:

(vii) $Ix \to x \alpha x \,.\, (z)(z \alpha x \to z = x)$

The notion of an *individual* is defined by

(viii) $I(x) \longleftrightarrow$ If $(\exists w)(x \alpha w)$

It is clear at once that the α-relation is also quite different from the π-relation of the Noll-Suppes theory of bodies. Where, $A \,\pi\, A$ holds, $a \,\alpha\, a$ does not (for A a body and a an aggregate). What is more, not all parts of an aggregate are components of it. The transitivity principle, if $x \,\alpha\, y$ and $y \,\alpha\, z$, then $x \,\alpha\, z$, does not obtain where y and z are aggregates. However, transitivity is provable in case x and y are just the same individual. In fact, given the joint assumption of $x \,\alpha\, y$ and $y \,\alpha\, z$, it *follows* that $x = y$, and therefrom that $x \,\alpha\, z$. So aggregates are not mereological classes or Suppesian bodies. Aggregates are entities that suggest themselves for the analysis of idioms in which there are plural constructions and mass terms, e.g. for such expressions as "the stars that presently make up the Pleiades galactic construction." Unlike sets, aggregates are physical entities in space-time, capable of action and of change, and susceptible of coming into and going out of existence. Unlike mereological classes and Suppesian bodies, not all parts of an aggregate are parts that *make up* the aggregate; no aggregate is its own member-component; and no aggregate may be the component member of any aggregate (i.e. aggregates are always aggregates of individuals). By virtue of the first of these three differences, one might be able to think of an iron chain as an aggregate, the ironness of which divides over its parts (i.e. its member-components, i.e. its links), and not worry about, e.g. the atomic parts of these components. They are parts that do not matter. And by virtue of the third of these three differences, aggregate theory avoids Russell-Zermelo problems of aggregates of all aggregates that are not component-members of themselves. In Burge's theory, predicates of aggregates are kept reasonably well-behaved and projectible. Aggregate-theory is rather well-suited, therefore, to our interest in composition and division. How, then, should the part-whole relation be taken for the purposes of a reasonable analytic understanding of the fallacies of composition and division? Not, certainly, as set theoretic membership or set-theoretic inclusion. And not as Suppesian body-parthood. The better prospect, though I do not claim perfection for it, is Burgean componentiation. In this fashion, a formal theory—the formal theory of aggregates—seems next-to-indispensable for a decent account of the traditional fallacies of composition and division.[22]

Footnotes

[1] Saul Kripke, "Semantical Analysis of Intuitionistic Logic I", **Formal Systems and Recursive Functions**, Ed. by J.N. Crossley and M.A.E. Dummett, Amsterdam: North-Holland, 1965, pp. 92-130; and Jaakko Hintikka, "The Logic of Information-Seeking Dialogues: A Model", undated typescript. To be clear about this point, it is not in fact the case, so far as I know, that there exists any theory of circularity that qualifies for the status of a mathematical system. Fragments of such a theory exist which involve the use of Kripke's intuitionistic logic (which is a mathematical system) and also Hintikka's dialogical system (which is not a mathematical system).

[2] Hao Wang, "On Formalization", in Copi and Gould **Contemporary Philosophical Logic**, New York: St. Martin's Press, 1978, pp. 2-13.

[3] Wang, ibid.

[4] The interested reader might wish to consult some of the following: "*Argumentum ad Verecundiam*", **Philosophy and Rhetoric**, 7, 1974, 135-53; "On Fallacies", **The Journal of Critical Analysis**, V, 1974, 103-11; "*Petitio Principii*", **Synthese**, 31, 1975, 107-28; "*Ad Baculum*", **Graser Philosophische Studien**, Vol. 2, 1976, 133-40; "*Post Hoc, Ergo Propter Hoc*", **Review of Metaphysics**, Vol XXX, No. 4, 1977, 569-94; "*Petitio* and Relevant Many-Premissed Arguments", **Logique et Analyse**, Vol 77-78, 1977, 97-110; "Arresting Circles in Formal Dialogues", **Journal of Philosophical Logic**, 7, 1978, 73-90; "Toward a Theory of Argument", **Metaphilosophy**, Vol. 8, 1977, 299-315; "Laws of Thought and Epistemic Proofs", **Idealistic Studies**, to appear; "Circular Demonstration and Geach/von Wright Entailment", **Notre Dame Journal of Formal Logic**, to appear; "Composition and Divi-

sion", ***Studia Logica***, Vol XXXVI, 1977, 381-406; *"Ad Ignorantiam"*, **Dialectica**, to appear; "Equivocation and Practical Logic", ***Ratio***, to appear; *"Ad Hominem"*, to appear.

[5]See "Arresting Circles in Formal Dialogues", pp. 82-89.

[6]See *Post Hoc, Ergo Propter Hoc*, pp. 574-577 and 591-593.

[7]Richard Routley, "The Implication Connection, and the Ensuing Inadequacy of Irrelevant Logics Such as Classical and Modal Logics", unpublished typescript, 1978. In this paper Routley develops a system in which contradictions can be true yet in which not everything is true. In other words, in such a system, contradictions do not make for theoretical psychosis. Routley furnishes an argument intended to show that the world is in fact inconsistent. This metaphysical claim not only provides the motivation for a logic in which such contradictions might obtain, but it is also of very direct relevance and importance for any analysis of the kind of *ad hominem* argument in which there is a charge of inconsistency.

[8]See, for example ***The Logic of Questions and Answers***, Nuel D. Belnap, Jr., and Thomas B. Steel, Jr., New Haven and London: Yale University press, 1976; L. Aqvist, ***A New Approach to the Logical Theory of Interrogatives***, Uppsala: Almqvist & Wiksell, 1965; D. Harrah, ***Communication: A Logical Model***, Cambridge, Mass.: M.I.T. Press, 1963; J.M.O. Wheatley, "Deliberative Questions", *Analysis*, 15, 49-60; Douglas Walton, "The Fallacy of Many Questions", unpublished typescript, 1978.

[9]"Arresting Circles in Formal Dialogues", pp. 83-85.

[10]Tyler Burge, "A Theory of Aggregates", *Nous*, XI, No. 2, 1977, 97-118; and Woods and Walton, "Composition and Division".

[11]George Berger, "Temporally Symmetric Causal Relations in Minkowski Space-Time", *Space, Time and Geometry*, Ed. Patrick Suppes, Dortrecht: Reidel, 1973, pp. 56-71; and *"Post Hoc, Ergo Propter Hoc"*, p. 581.

[12]For a discussion of the distinction between inference and entailment rules the reader could consult Woods and Walton "On Fallacies", see also *"Petitio Principii"*, pp. 112-13 and 125-26.

[13]"Composition and Division", pp. 388-396. See also the Appendix to the present paper.

[14]See for example, "Toward a Theory of Argument", "On Fallacies", *"Petitio Principii"*, and "Arresting Circles in Formal Dialogues".

[15]Waiving for now discussion of an interesting thesis that one tends to associate with Hilary Putnam, namely, that logical theories are also empirical.

[16]Of course, various bits and pieces may yield up quite convincing axiomatizations, but that is a different point.

[17]I have appropriated most of this list from Baruch A. Brody's article, "Logical Terms, Glossary of", ***The Encyclopedia of Philosophy***, Vol. 5-6, New York: Macmillan Publishing Co. Inc. & The Free Press, 1976, reprint edition 1972, p. 64.

[18] S. Léśniewski, "O podstawach matematyki (On the Foundations of Mathematics)", *Przeglad Filozoficzny*, Vol 30 (1927), 164-206; Vol. 31 (1928) 261-91; Vol. 32 (1929), 60-101; Vol. 33 (1930), 77-105; and Vol. 34 (1931), 142-70.

[19] Patrick Suppes, "Problems in the Philosophy of Space and Time", in *Space, Time and Geometry*, ed. P. Suppes (Dordrecht: Reidel Publishing Co., 1973). See especially pp. 392-95.

[20] T. Burge, "A Theory of Aggregates", *Nous*, XI, 1977, 97-118.

[21] For a more complete examination of the suitability of the formal theory of aggregates to the analysing of composition and division, the reader may wish to consult "Composition and Division".

ARGUMENTS THAT AREN'T ARGUMENTS

Peter A. Minkus
York University

I. INFERENCE PRINCIPLES IN CONTRAST TO THE UNDERMINING OF CONFUSIONS

I distinguish three senses of 'principle of inference': (1) a formal law, (2) a habit of inference, (3) a habit of inference that isn't confused.

Only where "principle of inference" is used in the third sense can it be truly said that principles of inference don't mislead, or don't confuse. (1) Formal laws are principles applied in too many cases and are invitations to distortion. They make out, to give an example, that the French orator who says, "All Franchmen love wine, but my followers, French and of the elite, don't" was contradicting himself, or not understanding what he said, or improperly using language, etc. (2) Habits of inference are similar in this respect. For example, a person might insist that of parts you must say they are smaller than wholes and so, when his son blows a soap-bubble form which part divides and becomes bigger than the original, he might correct his son who says "There goes a part bigger than the whole". What pointlessness, what misled pedantry to insist on such principles in such ways. And sometimes such principles and such insistence amount to vast trouble. This is what Ludwig Wittgenstein had in mind when he spoke against the person who argues, "If there can be some mistakes in speaking language, then it can be that there should be nothing but mistakes in speaking language" (cf., ***Philosophical Investigations***); also, compare Aristotle on "All S in P" entailing "Some S in P".

If you want to be sure that your are facing a case of (3) a habit of inference being followed in a way that is not confused, you need to look beyond the form-of-the-expression-used ot the context, and beyond the given context to the other cases in which the user uses the principle. If you find the user allows for different sorts of 'all' in the contexts in which the use of such different sorts of 'all' is established and not misleading and useful, you will accept from him a principle that 'all' both allows and doesn't allow exceptions depending on the case, or a principle that 'all' doesn't here and there allow exceptions, depending on the case. Such a user is liable not to be pedantically pointless about the French orator, and is liable to understand and maybe describe why different kinds of 'all' differ. There are 'all's' of hurried emphasis, and 'all's' of careful formulation.

Similarly for arguments from the stick (*ad baculum*) and arguments from authority, etc. In different places they will call for different descriptions. A father (or a bishop), to encourage courage, appeals to what he (or some saint) said some time ago, and even if he is unable to specify when and where the authority said it, he expresses the kind of reason for obedience that we encourage taking seriously as a reason. On the other hand, if a history student appeals to what was said by some Greek historian, and if he is unable to specify which historian, we hold it against him. Perhaps the date of the father's utterance couldn't be specified and perhaps there is no way of picking out the saint. Nevertheless we recognize these habits. And the case of the history student is different: for reasons inherent in the context we want to know which historian said the thing in question. Moreover the father's or bishop's, and the student examiner's, habits are all free of confusion here. (We insist on some information in the latter case, and on an attitude to certain kinds of utterances in the former. We are not illicitly passing to the wrong conclusion or misunderstanding what the premise supports, in either case.)

II. PLACES FOR PRACTICE IN KEEPING WITH PRINCIPLES AND GOOD ADVICE ABOUT SUCH PLACES

We expect scientists, at least in writing, not to confuse 'necessary' and 'sufficient', or 'is incompatible with' and 'is contradictory of'. And the principles we teach them for such places also matter in other places, for example in attacking one's adversary, where in confusion a person passes to a conclusion in violation of such principles—provided the context calls for resistance and correction, and provided there is particular harm done by the confusion. But there is also the case in which a man says "if he has an umbrella, that's a sign of good news. Oh dear, he has no umbrella, what a pity". In this case we could accuse him of confusion, and yet it also would be perfectly normal to accept his performance. Therefore it is part of good advice about teaching the more pedantic varieties of our language to insist that they are at home in just those contexts which call for pedantic varieties. So too it is part of good advice about teaching reasoning in general to insist that different varieties link different patterns of premiss-and-conclusion, and that as you attend in the right places with the right questions in mind you are liable to pick up the ways of the Ways of Reason. The Detective's Reason—is different from the Healer's Reason and both are different from those didactic and scrutinized contexts for which we train with distinctions between words in formal logic. Let us then *say* we teach for these didactic and scrutinized contexts where we so teach.

III. CONSIDERING PLACES FOR PERFORMANCES OF SPEAKERS IN CONTRAST TO CONSIDERING PRINCIPLES

I say to a colonist 'You are a colonist, therefore you are, coo, a looney", I hardly aim at an inference. I may aim to joke, or to insult, and you might be able to tell which. There are questions that can be raised as to what I am trying to do with my utterance, e.g., whether I am trying to assert myself, or whether I am well advised to assert myself, or whether it was my best plan so to assert myself.

In discussing arguments that aren't arguments, like the argument from authority and the *argumentum ad hominem*, such questions also arise. And if you make a habit of pursuing such quesitons, you will be able to recall the place in language of such arguments without being confined to the mere purist comment *'non-sequitur'*. In many places, it is of considerable use not to be tied to this comment, but to understand the place and the performance, its aim and its cleverness. This may, for example, be of great consequence to public speakers, advertisers, and pedagogues.

People aptly using the argument from authority, the argument from self-evidence, or the *argumentum ad hominem*, in sales-talk, or in political persuasion, don't show they have mastered a pattern of logical consequence but they show they manage appeals to *certain staple considerations apt to incline the audience*. There is no reason to accuse them of confusion unless they pretend to be pleading a pattern of logical consequence or unless the expectations there normally are of the audience are such that it is confusing for the audience to accept anything but logical consequence, etc.

If we nevertheless criticize such arguments as violations of deductive principles we should make it quite clear that no obedience to such principles is liable to be aimed at, that, for example, a salesman short of time and sure of the popularity with his client of Greta Garbo, might be able to do no better than, "Greta Garbo whom you adore wore these, won't you try them, Madame?"

I look on the Argument from the Stick (*ad baculum*) and the Argument from Authority as fairly familiar ingredients in playing well-known parts, for instance the part of being a *hero for worship by his adherents,* and the part of being *a master of disciplined men who is in a hurry to score*. When you can expect to be worshipped and can think of no better plan for getting on with your political ends than to stand up and make a bid for renewed acceptance, the

Argument from Authority, your own authority, falls into place. Similarly for powerful task-masters, generals, captains of ships and the Argument from the Stick (cf. Napoleon in *The World's Great Speakers*, ed. Copeland and Lamm, p. 85 ff. and note how close the text gets to such an argument).

In the jolly-good-fellow act a new proposal may be well defended by the *argumentum ad hominem*. In appealing to an audience with common acceptances or with a shared faith, a set of premises becomes available for the argument from self-evidence. There is nothing to stop us renewing the contact with context in even greater detail and thus reviving our awareness of language in public life for the benefit of those whose good advice depends on such awareness.

IV. PANARCHIC MODELS OF INFERENCE-STEPS

When it seemed necessary to build mathematics out of logic, a new "language", a new set of "games" arose, and the novelty was very visibly new. Yet when it seems necessary to tell the intending scientist that 'p is necessary for q' entails 'no q without p', it is not so clear to what extent and how this encourages a performance not in fact adhered to by scientists themselves. Consider an example. In common contexts there is no recovery without oxygen and that may be enough for a doctor or pharmacologist or physiologist in such a context to say "oxygen is necessary for recovery". But quite consistently he may add, "of course you can do without oxygen, if enough adrenalin is available".

It is often thought that all of the best places in philosophy and in academic life generally are pedantic, for scrutiny of one's steps in the light of the principles that govern the best practice. This is a wretched model, which tends to prevent philosophy and other disciplines from reaching truth and defeating error in natural language. It puts a cramp on practitioners, preventing them from being aware of our language, of common sense, and of the world picture.

Panarchic models have reduced practitioners to G. E. Moore's way, to the *Principia Mathematica* way; earlier practitioners were reduced by them to Aristotle's way. Panarchism is connected with a vast silhouette of families of confusions in the history of logic. [At one stage all the best logicians practiced for all steps to become mathematical steps (cf. Poretsky in Styashkin).]

There is a great difference between observing an intellectually appealing method, whether it is intellectually cramped or not, on the one hand, and restoring the sense of our practice with reasoning language when we are not confused, on the other.

V. METAPHYSICAL ROOTS OF TROUBLE CONNECTED WITH THE PRECEDING

We defend principles by appealing to facts and have no defence but cases, unless we believe in miraculous intuitions, super-plans, etc. Yet we feel there is more to principles than the cases in which they make sense, the facts of use from which we defend them. Again and again we get this impression with formal principles, and then at the very least we find we should allow exceptions (e.g. for "$p \rightarrow Mp$", where its use is uncomfortable through the insight that logical truths can't be guaranteed by contingency). Facing exceptions we sometimes abandon the principles, but often we keep them and formulate further principles to cover the exceptions (cf. "p" means "p makes sense" here). We make further principles instead of recalling that each principle carries weight only where it makes sense. The Rails to Infinity seem laid for us, because we forget that our rules are bids for obedience not wearing in their faces their limited defence from particular cases. We are trapped in this manner in formal and informal logic alike. We hurry, for example, to establish vigour throughout the dictionary and become tyrants at the expense of language-variety, forgetting that rules only have the value of their supporting cases.

VI. WANTING NEW HABITS OF PEDANTRY VS. PRINCIPLES
(sense 3)

If I don't want the argument from authority used in fields where there is difference of opinion and bias I may serve one set of purposes well and yet by the generality of my principle, undermine another. We are used to appealing to authority as parents, as priests, or as masters of discipline, and here, without confusion, pass from bias and opinion in the premiss to the same bias and opinion in the conclusion. Taking the rule not to argue from authority in fields of bias and opinion seriously, won't we be less equipped to encourage obedience and discipline, and mightn't we incline to call our opinions and our bias "scientific insights" or something of the sort?

If we look at the *Given*, at what happens when authority is appealed to, describe this instead of looking for new principles, then we needn't encourage confusion. Mere description can be in keeping with established and unconfusing principles. The matter is merely one of adjusting our descriptions for different audiences in order to avoid any miscomparisons the audience might have.

The question of new principles, and where and how they are useful without becoming alienating, is a matter of the structure of good advice and how such structure differs for different performers in different contexts. This needs facing in detail. What is an advantage, a useful increase of caution, in preparing a work on the history of Canada, may be quite out of place if you deliver an address as a hero-to-be-worshipped.

VII. WHY THE NORMATIVE DISCIPLINES ARE EVEN MORE TROUBLESOME THAN SAID SO FAR

What we tend to do in logic, grammer and other rule-pronouncing activities is to be fond of rules without qualifications and also—which is the new consideration here—of an impressive manner in collecting, putting together, and articulating these rules. Thus we borrow from impressive activities (such as mathematics, philosophy) considerations of rigor, impartiality, etc. The combined intellectual and efficiency ideals we fall for remove the formulations yet further from language. The species of Panarchic Models that govern collecting, putting together and articulating logical principles provide further obstacles to anyone concerned to understand how principles *can* and *can't* work. We provide a quasi-mathematical or quasi-philosophical contexts for principles that are not used as parts of a calculus or through philosophic support, but that have limited *corrective and supportive* uses supported by limited ordinary cases. We make it appear that these principles constitute a complex mathematical apparatus of unlimited use or that they are the results of profound arguments (here recall again Poretsky in Styashkin's book, or Aristotle's philosophising on the nature of *episteme*). These special forms of method-worship remove us further than ever from a realization that we require principles with cautions as to their place of use and recall of cases they are supported by. The fiction that anybody's good advice necessarily limits their language to a special, favourite variety of language which is kept trimmed and hedged to comply with panarchic models both of principles of inference and methods of collecting, presenting and connecting principles of inference, has to be undermined by looking at the things we say when we are not tyrannized by a fashion in logic.

POST-SCRIPT

Out of the discussion following the reading of this paper at the Symposium on Informal Logic, June 1978, Windsor, Ontario, I would like to recall three points and reply to them:

1. Somebody, as if to contradict what I had said of the French orator example, produced a version of the Epimenides conundrum; in it a Frenchman was supposed to have said, "The French are liars" or something of the sort and his remark would be treated by me as self-defeating. Wouldn't I be unduly normative?

In answer to this, I have to say that in getting rid of the puzzle, in philosophizing, quite often, it is a step to call a puzzling utterance a form of nonsense. This is a step in the right direction, since you may help a philosophizing hearer puzzled by an expression if you undermine the puzzling expression through contrasting it with expressions that have recognizable purposes. In so undermining the puzzling expression you may say something like "But p is self-defeating", "p has no rules of use" and you may then look to people who are new to your activities to be using normative, over-ambitious, context-disregarding principles; you may look to outsiders to be setting up a kind of *a priori* prison. If, however, they bother to see that you are trying to undermine confusions without necessarily being pedantic in how you insult the confusions, they will admit that you are pursuing that principal polarity, that overcoming of puzzles, which makes clarification divine by contrast to traditional philosophy; they will then see, in Wittgenstein's philosophy, not a collection of new raving principles, but the release from obscurity deserved by man.

2. Another participant in the discussion produced, as if to contradict me, the case of a psychiatrist whose *non-sequitur* in dealing with a patient brought the patient to an unhappy end.

I had said, in the discussion, that considerations we are liable to pick up in acquiring traditional academic habits of argument wouldn't help us, by and large, in getting on. And I must admit that in a range of cases, namely where we face experts and professional men and we can make out that they *deductively* over-infer from premises, we can rely on the sort of considerations we use in saying "that doesn't follow". But these cases do require, *ex hypothesi*, that somebody looks to *over-infer deductively* as opposed to inferring with the confidence of *mere sanity or experience*. They require some indication that he thinks he has got hold of the only proper and pertinent variety of language, or some super-insight into abstract relations. Such indications we don't by and large find in our communicative encounters with traders, neighbours, family and bureaucrats. We find them where there is the prestige of an expert who handles a "preferred variety of language" or where there is intellectual or other verbal obstinacy. If you like to say that we find these often enough to be prepared for them also, I will agree, reminding you that the considerations pertinent to "it doesn't follow" include indications as to the use of expressions then and there, and are therefore not context-disregarding abstract principles.

3. Somebody said, "The point is that there is one and only one right way of inferring and we know it". I called that a model. There are principles propounded by logicians or philosophers and at times we think everyone will go by them. If we mean such principles by "one and only one right way of inferring", we are misled. There is nothing about the fact of logicians or philosophers propounding the principles which would guarantee they don't confuse, mislead, etc. The fact is we suffer a great accumulation of principles and if we don't get the habit of taking a distance from them, we won't be able to deal with puzzles. Take for example the puzzle discussed

near the end of Moore's *Commonplace Book*, of a "syllogism" proving that you may infer a necessary conclusion from a necessary and a contingent premiss. The "syllogism" was,

all brothers are siblings
all siblings are brothers
all brothers are brothers.

You may undermine this puzzle by taking a distance from the assumption that whatever formally is a case of the syllogism in Barbara, makes sense as an inference. If somebody were to "argue", in conditions of market-trouble,

all steel screws are small metal artefacts
all small metal artefacts are steel screws
all steel screws are steel screws

you wouldn't understand at all the purpose of his performance. If nevertheless you hold on to Barbara and find yourself with too many principles and a clash of principles and puzzles, you experience one of the many ways in which logicians' principles (and philosophers' principles) turn out to be confusion, unbending, calculus-like distortions of language.

Footnotes

[1] This paper is dedicated to Ludwig Wittgenstein, *vir maximi ingenii*, the God of philosophy, philosophers and the present text.

PEDAGOGY AND PRACTICE

CAN THE ABILITY TO REASON WELL BE TAUGHT?

Robert W. Binkley
The University of Western Ontario

My title, "Can the ability to reason well be taught?" is a question. It is not unlike a question about virtue which Meno once asked Socrates, a question to which, as is well known, he never really did get a straightforward answer. Some in my audience, familiar with this precedent, may be fearful of receiving a similar Socratic runaround from me, so I hasten to allay this anxiety by giving a definite answer at the outset. It is Yes, the ability to reason well *can* be taught.

And a good thing, too, for several reasons. First, of course, the ability to reason well is itself a good thing, and if we can make its incidence in the population more frequent by teaching, well, that will be a good thing too. But this is the argument to be used to persuade taxpayers to continue to send their youth to us and to pay our salaries, and while that is important, I want to focus here on a second reason, one more directly relevant to us as participants in a conference on informal logic. For I believe that a major motivation for the recent growth of interest in informal logic, and I suspect also a motivation behind this conference, is first the belief that the ability to reason well can indeed be taught—and further, that it can and should be taught by teaching informal logic. What a pity for us and our conference if these beliefs should prove false.

Notice that this motivation has two components; the belief that reasoning ability can be taught *somehow*, which belief I regard as true; and the belief that it can only be done, or can best be done, by teaching something known as informal logic. About this second component I am somewhat ambivalent, and my doubts stem only partly from uncertainty as to what exactly informal logic is.

My question is essentially an empirical one, and so my affirmative answer to it ought to be backed up by empirical evidence. Unfortunately, I do not have such evidence to offer. We can easily imagine the sorts of controlled experiments that could be devised to provide that evidence; essentially, measure reasoning ability at the beginning of the course, and again at the end, with a control group of students who don't take the course. But I have done no experiments of that type. Nor am I very familiar with the literature, not large, of those who have worked in this area. Such evidence as I have just barely approaches the level of the anecdotal, and is simply the result of my own experience in trying to teach this ability by various methods.

That we do not have a large body of empirical data on this point is not surprising. There is, of course, the problem of measuring the ability to reason. But more important is the fact that the teaching of this ability, at least the kind of teaching I have in mind, is largely done by philosophers, and philosophers, even empiricists, are not much given, by temperament or training, to actually performing experiments.

So I shall not present my answer as established fact; I shall treat it instead as an article of faith. As such, I believe, it will not be without some utility. Another contributor to this conference has observed that self-deception can sometimes have survival value, and it is possible that this is an instance of that phenomenon. Most of us, I imagine, have tried on occasion to do something in the way of teaching this ability, and have sometimes felt that our efforts were not achieving the results desired. Very naturally, then, though at the risk of committing the fallacy in modal logic of reasoning from *didn't* to *couldn't*, we ask, "Can the thing be done at all?" My purpose in dilating upon this article of faith, is to address this moment of pedagogical despair. I am not really going to say anything new; I will just try to say some of the things that most of us feel when involved in this kind of teaching. Perhaps this will have a group therapy effect.

I said a moment ago that the current interest in informal logic is in part due to the belief that it is the only or best way to teach the ability to reason. However, the two things must be kept distinct. It is one thing to set up improved reasoning as one's primary goal in teaching; it is another to decide to teach informal logic, as the means to this end. The two decisions frequently occur together, but the relationship is a contingent one. One way of characterising it is to go back to the kind of course that used to be known in the trade as 'baby logic'. This course, in my experience at least, was an introductory course in logical theory. *Ideally*, one would present a rigorous treatment of first order logic, with meta-proofs of soundness and completeness and so on. But this ideal could never actually be achieved because the first year students, to whom in the main the course was directed, had, most of them, neither the interest nor the ability to accomplish that, nor was there enough time. So a compromise was made, and as much first order logic was crammed into their heads as was possible, but this wasn't much, and it was soon forgotten. The course thus became an abbreviated and watered down presentation of symbolic logic. That is why it was called *baby* logic; it was a weak imitation of the real thing, like the toy carpenter's tools which are sometimes inflicted upon small children.

What was the purpose of this course? Here, I think, there was some confusion and the purposes were often cross. The students took the course because they wanted to improve their reasoning ability. Or they took it because it met a requirement, but it did so only because at some previous time some committee of professors had been under the impression that it would improve reasoning ability. (In this context, the course was often recommended as the moral equivalent of mathematics.) By this very natural understanding of the purpose of the course was often not shared by the instructor. The instructor would regard the science of logical theory as one of absorbing intellectual interest (which it is to the right sort of mind), and would want the students to appreciate that. Or, setting intellectual delights to one side, logical theory was regarded as a body of truth (which it is) that the instructor was eager to impart to those eager to learn. Or again it was thought that a grasp of symbolic logic is essential for an understanding of certain topics in philosophy (which is true), and this sometimes supplied the instructor's rationale for the course. That the course might contribute to improved reasoning was doubted by many instructors, who thought that this improvement was either impossible by teaching, or at any rate unnecessary at university level. Other instructors felt that perhaps some improvement in reasoning might come as a sort of by-product of the course. It was sometimes said, for example, that the course would give the student an example of what really rigorous thinking is like, and that this would do the student good (which of course is true up to a point).

I am not saying here that the instructors of the old baby logic course were wrong; they were just not even *trying* to provide what the students came to the course to get. Naturally, this led to a certain amount of frustration for all concerned.

This problem did not go unnoticed, and various moves were made. One of them—the one I want to consider—was to change the baby logic course by rethinking its purpose, and consciously and deliberately making improvement of reasoning ability its primary goal; other possible benefits—appreciation of structure of intellectual beauty, contact with an example of rigorous thought, etc. were reduced to the status of dispensible by-products, nice if you could get them, but not to get in the way of the main purpose. Sometimes this change of course goal was marked by a change of name, say, from "Introduction to Logic" to "Critical Reasoning".

Notice that what defines the new Critical Reasoning course, in the first instance, is its goal. The students want to have their reasoning abilities improved, and this course sets out to do that for them, using whatever means are suitable for the purpose. Nothing yet has been said about the role in this, if any, of informal logic. In my own view, the conscious and serious adopting of this goal by the instructor is the single most important step in any educational reform in this area. Once that end is fixed, the means in all their variety will tend to fall into place. So it will be worth a moment to inquire why instructors resist doing this. There are two main reasons, I think.

First, a course aimed directly at improving reasoning ability threatens to be boring to teach. One will not be dealing with interesting ideas as such, as one would, say, in an introduction to philosophy course. One will just be drilling away at some fairly elementary intellectual skills. The aim of the course will be to prepare the student to think about interesting things, but the course itself will not directly concern interesting things. It is rather like cleaning up the house in preparation for a party; dull work in itself, with the real fun to come later. At bottom, I think, this reason is sound. The critical reasoning course must be regarded as work, not play, by the instructor. He can hope for the teacher's reward of seeing the light dawn in the minds of some of his students, and some intellectual interest may be borrowed, so to speak, from the examples chosen to illustrate the methodological points. But the main work of the course should not be viewed as an exploration of interesting ideas; it is rather a matter of skill teaching, like a beginning French course, or a course in English composition. Philosophers are not used to teaching courses like that, and have a tendency to drift away from the practical work in hand to speculation about the more interesting basic principles underlying it. At least I do. In the present context, that is part of the problem.

The second main reason, I think, for instructor resistance to the critical reasoning course is the thought that it is somehow fradulent. There is evidently great demand for improved reasoning ability. Students want it, and parents and professors in other disciplines are eager that they should have it. It is rare indeed to find anyone who is satisfied that others, especially the young, already reason well enough. But can the critical reasoning course really deliver? This is the theme that I want now to pursue at some length. And it brings me back to my original question.

Virtue is also something that it is commonly thought should be instilled in youth, and demand for the teaching of it would be heavy if it were thought to be teachable. The teaching of it has been undertaken from time to time, but often too there has been the doubt that the enterprise might be fraudulent. It was, of course, in such a context that Meno addressed his famous question to Socrates, and their discussion of it, I believe, is not without a certain value for us.

When Meno asked his question, he was immediately tripped up by Socrates, who insisted upon knowing first what virtue is. I shall also have a bit to say about what this thing is that we are interested in teaching, but first I want to pin down the meaning of the other central word, "taught". As should be clear, I don't mean to consider the whole range of interpersonal transactions that might reasonably be included under Teaching. I mean to limit attention to what can be done by a philosopher-professor in a first year course in a Canadian or similar university—a largish course, one in which the professor is outnumbered by the students by, say, at least 50 to 1, and a course which has set itself the goal that I have specified for the critical reasoning course. That is, reasoning ability is to be taught directly, not as a by-product of other teaching. I shall also use 'teach' as an achievement word. A thing is not taught by a teacher until it is learned by the student.

So much for the moment about teaching; now something about the thing to be taught. I have been calling it the ability to reason, or the ability to reason well, but this is not entirely adequate. It goes wrong in suggesting that we are interested just in a particular skill, or bundle of skills, like the ability to play the piano. Of course, we *are* interested in a certain bundle of intellectual skills, but that is not the whole story. If I have the ability to play the piano, then, given a piano, I can sit down and actually play it when and if I want to. I can, so to speak, turn my piano playing on or off at will. But the goal in a critical reasoning course is not, or is not merely, to inculcate certain skills that permit related activities to be turned on or off at will. What we want in addition is an attitude, or set of attitudes, towards the use of these skills. In the case of piano playing, the relevant attitudes would perhaps be enjoyment of piano playing, eagerness to engage in it, interest in it, whether one's own playing or that of another, and so on. Perhaps we could sum this up under the label 'love of the piano'.

Analogously, the attitudes we seek to foster in the critical reasoning course might be summed up under the label 'love of reason'. We not only want our students to be *able* to reason well; we want them actually to do it, and so we want them to be eager to do it and to enjoy it—to think it important. We want it to assume an important place in their lives.

We need a term to refer to this combination of the ability to reason together with the love of it, and I think the term 'rationality' is well suited for the purpose, and will so use it. And I think it is fair to say that rationality, as I have roughly characterised it, is a virtue. I will be content on this occasion to treat it as an intellectual virtue, though I think that really it is broader than that, and that it can and should inform the realm of conduct, and indeed even that of feeling, as well as the realm of pure thought.[1]

We must now notice that if rationality is a virtue and we are trying to teach it, then we are trying to teach a virtue, which brings us back to the unfortunate Meno. It is true that we are only concerned with one particular virtue, not that One Virtue, or Virtue Itself, which Meno's question concerned. But perhaps we need not explore that particular blind alley now, and this will spare us some of the embarrassments that Meno suffered in trying to get his definitions together.

Socrates and Meno make a key move in their investigation when they decide to use the method of hypothesis, and in particular, to employ the hypothesis that virtue is knowledge. It seems evident to them that knowledge and only knowledge can be taught (87b), and this leads them to affirm that virtue can be taught if and only if virtue is knowledge, and then to inquire whether it is.

Their first conclusion in this investigation is that virtue is indeed knowledge. Their thought seems to be this: virtue is an attribute of soul that always produces proper guidance in the conduct of life, but knowledge is the only thing that always produces proper guidance, hence, virtue is knowledge, and so can be taught.

At a later stage, however, they are obliged to reject this result in view of the fact that there do not seem actually to be any teachers of virtue, which implies that those who have it must get it some other way. And then they discover the flaw in their reasoning. Knowledge is not after all the *only* thing that produces proper guidance in life; there is also right opinion, which does the job very well for those who have it. Right opinion, of course, is not the same as knowledge, and is not taught; it is hard to see where it comes from, and indeed it would appear to be a sort of inexplicable gift from the gods. So, they conclude at last, virtue is right opinion and cannot be taught, though Socrates does feel that a more definitive resolution of the problem might be achieved if they could only discover what virtue essentially is.

Now all of this, it seems to me, has important application to our particular virtue of rationality. Let us return to the moment at which the two had convinced themselves that virtue is knowledge, and so can be taught. Knowledge of what? we want to ask. Or more particularly, if rationality is knowledge, what might it be knowledge of? Here, it seems to me, the most plausible candidate is logical theory. What else is there in this area, after all, to be *known*? So let us explore this idea.

What exactly *is* logical theory? Or rather, what *roughly* is logical theory, for I shall try to make this long story very short. Theory, it seems to me, is an intellectual structure designed to give understanding. Logic is the study of reasoning, as correct or incorrect. Logical theory, then, aims at the understanding of correct and incorrect reasoning. More exactly, the goal of logical theory is achieved when we understand *why* certain reasoning is correct and certain other reasoning is incorrect.

Logical theory divides into two parts—formal and informal. Formal logic considers those aspects of correct and incorrect reasoning that depend on logical form alone. Informal logic considers everything else. Formal logic now exists as a sophisticated science, expressible in its own symbolic language. Informal logic, at present, and for the most part, is not a sophisiticated science at all, but rather hovers on the boundary between science and thoughtful common sense. Both parts of logical theory, presumably, will be relevant to the virtue of rationality.

To know logical theory, then, is to possess an intellectual structure which explains why reasoning is correct or incorrect, and perhaps also why incorrect reasoning occurs. But if this is the knowledge to be associated with the virtue of rationality, then I think Meno and Socrates would be right to point out that it does a certain job, roughly, provide proper guidance to thought processes, a job which can also be done by something else, something which they would call *right opinion*. For I think we may identify right opinion, in the context of rationality, with what is sometimes called logical instinct—the intuitive sense of what is correct or incorrect in reasoning. The person with such an instinct reasons correctly even though ignorant of logical theory. This person will not know, or be able to explain, *why* the reasoning is correct or incorrect and will be comparable to Pericles, who manifested great virtue in the conduct of Athenian affairs, but was unable to teach virtue, even to his sons.

This presents us with a problem. If correct reasoning can proceed both from knowledge, that is, grasp of logical theory, and also from right opinion, that is, intuitive logical sense, then the question arises for the teacher of reasoning which of these ought to be taught. Or should both be taught? I shall answer *both*, but a few points must be cleared up first.

The question, we should notice, could not arise for Socrates and Meno because of their conviction that only *knowledge* can be taught. I don't want to disagree with them about this, but instead point out that I am using 'teach' in a rather special sense—roughly, teaching is whatever you can get away with in a course of the kind I am considering; about *that* I don't think Socrates and Meno had any settled opinion. Thus in my sense of the word, it is an open question whether right opinion can be taught. I *think* the answer is Yes. That is, I think that there are things you can do in such a course the effect of which will be that students have better logical instincts. That, indeed, is a main part of the reason why I say that rationality can be taught.

A second point about knowledge and right opinion is this. For various reasons we moderns will be tempted to identify *right opinion* with *true belief*, and then to think of knowledge as some species of true belief, true belief that is justified, for example, or that has been awarded some other of the epistemologist's gold medals. Thinking in these terms, we would say that right opinion is automatically taught if we teach knowledge since knowledge just *is* a kind of right opinion. What must be pointed out is that this is not the way we are using these terms here.

Right opinion, as conceived here, is not simply true belief about the correctness or incorrectness of certain reasoning; it is such a belief arising out of an intuitive sense or knack. For a person with opinion in this area, certain reasoning 'feels right' and other reasoning 'feels wrong', and if the opinion is right, these feelings will coincide with the logical facts of the case.

A third point is that the right opinion considered here concerns *particular* cases of reasoning; it recognizes *this* argument as correct and *that* as incorrect, but it does not generalise or consider things abstractly.

In these respects, right opinion is quite different from knowledge. Knowledge, here, is understanding why certain *kinds* of reasoning are correct or incorrect. Knowledge deals in generalities; categories of reasoning—logical forms, we might say—it declares this inference to be valid *because it is of such and such a kind*. Thus it is not primarily about particular cases. And of course, as a kind of understanding, knowledge must be different from any instinctive sense.

Knowledge, for Socrates and Meno, is not only different from right opinion; it is superior to it. Right opinion is like the clever statue made by Daedelus. It is a fine thing to possess, but you must chain it down or it will run away. Right opinion too will run away, that is, will succumb to the blandishments of sophistry or passion, unless it is chained down. And what chains it down is knowledge, that is, understanding of the reason why. I think that this is true, profoundly true. The intuitive logical sense succumbs all too easily; knowledge of logical theory provides some protection. And since the protection of sound logical instinct against the temptations of sophistry and passion is certainly one of the goals of our teaching—is a component of rationality—the case for teaching knowledge—logical theory—is easily made. Of course, there remains the question which parts of logical theory have top priority. About this something later.

The case for teaching right opinion is different. Two points need to be made. First, while it would have seemed evident to Socrates and Meno that knowledge, if one could only get it, would suffice for the task of guidance, I think this would be a mistake. Guidance is required in particular cases but knowledge deals in generalities. To give guidance, knowledge must be applied to the particular case, and this requires the ability to perceive the particular case as falling under one of the categories dealt with by knowledge. You must perceive an argument as having, say, the form of *modus ponens* before you can bring your knowledge of logical theory to bear upon it.

This ability to perceive cases under the categories of logical theory is, I believe, a form of right opinion, and if so, then at least that much right opinion must be added to knowledge if the desired result of rationality is to be produced.

But I believe that still more is needed. For one thing, most of the time, it seems to me, reasoning proceeds and ought to proceed on the level of opinion—logical instinct. There is not time for anything else. Knowledge—the understanding why—should only be brought in when the going gets tough in some respect. So I would hold that right opinion—sound logical intuition—is in general a crucial component of rationality.

And there is one further point about this. There is an important sense, I believe, in which knowledge of logical theory rests upon true opinion. The knowledge we are considering gives us understanding of why certain reasoning is correct or incorrect, but if we ask how we gain this understanding, I think we must say that it is because we are persuaded by the various arguments offered by logicians, arguments which we perceive instinctively to be correct. Without some minimum of logical instinct, we could never come to possess logical knowledge in the first place.

Knowledge, then, is needed for combatting sophistry; true opinion is needed if knowledge is to be applied, and is also needed as a time-saver as well as a foundation upon which knowledge of logical theory can be built. But strictly, this only shows that these things are needed, not that they need to be taught, which is what I set out to prove. But if they are necessary for rationality, then if we are to teach rationality we must teach them, unless we can suppose that our students are already in possession of them.

Now I do not think it will take very much experience in a logic class to convince you that the first year student does not invariably possess a deep grasp of logical theory; he or she will generally not know why certain arguments are correct or incorrect; or at least, will not be able to *say* why. So if logical theory is to be got at all, it must be got by teaching.

The situation is less clear when it comes to right opinion. Of course *we*, who later became logicians and philosophers, had very well developed logical instincts when we were in first year university, we loved to argue, and so on, and it is easy for us to imagine that the same is true of the students who now confront us. But this would be a mistake for several reasons. For one thing, most of our students, though thankfully not all, are dumber than we were or are. I mean with respect to the reasoning abilities here in question. This is mainly a matter of the selection process by which we selected ourselves into our present profession, but perhaps also a function of the selection process which now gives us our students. Second, our students differ widely among themselves in their reasoning skills. This is an area in which some empirical research would be very useful, but at least the grosser differences are evident to any instructor. Some students can't reason well, just as some students can't read or write well—even at the intuitive level. A third point is that the well developed logical intuition which we possessed in our youth may just possibly not actually have been quite as fully developed as we recollect it to have been.

The point I am after here is that we cannot count on our students already possessing satisfactory logical instincts, and if they do not, then it becomes our job to teach them, and so this must be included in the critical reasoning course.

Perhaps a word should be said at this point about who should take this course. Not all students by any means. Many of the brighter students already possess adequate reasoning skills which will be sharpened as their studies progress in other subjects. Often when I spot such students in time, I advise them to take introduction to philosophy instead. There are many other students who would benefit from a good critical reasoning course, but who have more important other needs, and who may hope to pick up reasoning skills in other ways and places. I think, in fact, that there may be some truth in the charge that it is really the poorer student who stands in greatest need of the critical reasoning course. Or perhaps one should say that a course adequate to those needs will not be of all that much use to the more able student. Sometimes in my blacker moods I refer to the critical reasoning course as a course in remedial thinking, a name which, I think, is not entirely without point.

But however that may be, we have reached the result that the students who turn up in the critical reasoning course are to be taught two things—knowledge and right opinion about reasoning. The next question is how this is to be done.

Meno and Socrates have little to say about the inculcation of right opinion; they seem to think that it comes by divine dispensation, and there is nothing very much that the university lecturer can hope to do about that. However, I think in this respect they were unduly pessimistic. You can do something about improving logical intuition; it is improved by drills and exercises. It does not seem practical to put one's students, like pigeons, into logical Skinner boxes, but I am sure that if one could, logical intuition would be found to improve. And there are a number of familiar classroom techniques that offer a more civilized approximation to this. The trick is to work out the right sequence of exercises, and to provide for motivation and the prevention of tedium. I think that this is really an important part of the critical reasoning course, one which we, or at least I, often shirk.

Socrates and Meno are rather more helpful when it comes to the teaching of knowledge. Gaining knowledge, they say, is a matter of recollection, and so the teaching of it must be a matter of reminding. This is very true, at least for the sort of knowledge that now concerns us. This is the understanding of why reasoning is correct or incorrect, and if it is to be of any value in resisting sophistry and so on, it must be developed out of the student's own thought. It must start with the student's appreciation of the correctness of some reasoning, which appreciation must then be extended to other and more complicated cases. The student must be satisfied, convinced of the correctness of the recommended logical procedures, and to produce this effect, the student must in effect be led to rediscover logical theory for himself. (Also, a bit of the Socratic sting-ray at the start to induce a suitable humility is often not out of place.)

Unfortunately, the teaching technique illustrated by Socrates for this purpose, one-on-one dialectic, is not cost effective in the context of the present day university. One wonders what he would have done with a class of 50 slave-boys. This is a real problem, but we have some techniques of our own that approximate to dialectic—class discussion, the skillfully programmed text, tutorial sections, and so on. The important thing to remember is that the goal in this part of the course is understanding, and that the student must be brought to see the light for himself.

This is a good place to introduce a comment about an aspect of rationality which we have allowed to drop out of sight in our concern for knowledge and right opinion, though it is very important. I mean the one referred to above as the love of reason. This too, I think, can be taught in our meaning of the word. Love is taught, in so far as it can be, by contagion and by pleasant association. This means first that the instructor must love reason, and must allow the students to see that this is so. It means second that the critical reasoning class must itself be a model, an enjoyable model, of rational procedure. Sweet reason and soft words must prevail there; egos must not be damaged, nor that of the instructor inflated. One of the great virtues of Michael Scriven's textbook is that in his discussion of teaching methods he points firmly in this direction. And a second point: there are some aspects of these courses about which you sometimes wonder why it has become the business of philosophers to teach them rather than, say, mathematicians or members of English departments. But in this respect, maintaining a classroom atmosphere of quiet and pleasurable respect for reason, and for each other, I think that philosophers are pre-eminently qualified.

Thus, to sum up this part of my remarks, the critical reasoning course has three aspects, the training of logical intuition by drilling, the pursuit of theoretical logical insight by some feasible substitute for Socratic dialect, and the whole to be conducted in such a manner that it will conduce to the love of reason.

I want now to make a few scattered suggestions about how such a course might be done, which will illustrate these ideas. I am not, I hasten to say, going to set out a complete plan. I don't have a complete plan—not one I'd recommend. Nor will I only be listing things that I have tried in the past that worked—for to some extent I will only be announcing good intentions for the future.

First, of course, there is the question of the textbook. The typical textbook, that is, what Johnson and Blair have called the Global textbook, will introduce the notion of argument and draw the distinction between deductive and inductive reasoning, and then move into propositional logic with truth-tables and perhaps deduction. Then we may have syllogisms with Venn circles, and the more ambitious text may launch some sort of attack on quantification. Typically there will also be some discussion of language and meaning, some mention of the Famous Fallacies and some sort of treatment of inductive reasoning. (Also, a few books do, and more of them should, have a discussion of decision theory, and other forms of reasoning to a conclusion about what to do.)

To my mind, this is not a bad sort of book to use, though supplementation with a New Wave or informal logic book is helpful. The important thing is to use the book in the right way, which means always in the service of the purpose of the course. For example, the concept of argument will have to be isolated and defined, along with related concepts, and the logician's regimented terminology introduced. (Some people object to this last step on the grounds that the terminology is no help in communicating with non-logicians, but I think this is a mistake. These terms pay their way in facilitating the formulation and comprehension of other important points.)

This grasp of the concept of argument is a point of logical theory; the student must be brought to understand what argument is. But there is also an instinct to be instilled about this. The student must get a feel for argument, must come to perceive discourse, where appropriate, as breaking up into a premise-conclusion structure. Here, it seems to me, the use of tree diagrams as introduced by Beardsley years ago, and used by Scriven and by Johnson and Blair and, in a more systematic way by Steven Thomas, is very useful. The point here is not so much to learn tree-drawing as a technique; it is to do it enough so that the student comes to feel argument structure. More generally, I think that this is where the New Wave makes its greatest and most needed contribution.

When we come to propositional logic, tension between the demands of knowledge and the demands of right opinion become more pronounced. The use of symbols seems to me to be justified from both points of view because of the help they provide in recognizing and discussing logical form. Here the most important thing is the use of symbols to abbreviate statements. The use of the connective symbols of standard propositional logic is less vital, but I think still useful. The discrepancies in meaning between these symbols and their conventional English counterparts present a certain pedagogical problem, but it is surmountable, and the results justify the effort; (*or* so I believe, *if* I may use a deviant 'or', not to mention the deviant 'if' that followed in the next clause—and what about that 'not'?) The contrast gives the student a clearer grasp of the meaning of the English words. There is also occasion for a little lesson on how words are the counters of wise men but the money of fools. More important, the logician's symbols, with their tie-in with truth-tables, facilitate for the student the fundamental insight that deductive correctness amounts to impossibility of counter-example. In the context of propositional logic, at least, we can *see* just what this means because the truth-table charts all the possibilities, and we can check them all out to see if there is a counter-example. This is important for logical theory, and is a good reason for including truth-tables in the course. But it is the only reason. Truth-table analysis should not be taught as a useful method of calculation, like long division, because it is not. It is rare indeed that one encounters a problem outside a logic book that is best solved by truth-tables.

What would be more useful in the sphere of propositional logic is deduction. This helps with the theoretical idea of chaining together simple obvious steps to bridge a long unobvious inference. But I think it also can provide useful training to the logical intuition in certain basic logical forms. (I assume, of course, a sensible and intuitive system of deduction. One does not want elegance here. Nor rigor, especially.)

Syllogism too can be used to make contributions both to theory and to intuition. Venn circles, it seems to me, do help to make an important theoretical point—that correctness of reasoning in this area has to do with the existence and non-existence of objects of certain kinds. But I do not think that the circles, any more than truth-tables, should be taught as useful methods of calculation. Nor do they help the intuition much. The traditional syllogistic rules are only a little better in this regard. I think what is most useful here for logical intuition are the immediate inferences. And they can help to make Aristotle's theoretical point—that a syllogism of one form can be transformed into an equivalent syllogism of another, and even that valid syllogisms can be 'reduced' to syllogisms of the first figure, though on the last point it is doubtful whether the classification of syllogisms by mood and figure is worth the trouble.

Skipping over quantification, as one ordinarily does in these courses, we come to language and meaning. Here I do not think there is any very pertinent theory to be taught. The true theory, some kind of scientific linguistics, is too complicated, and the easy approximations, such as the doctrine of connotation and denotation, are not particularly useful. What is needed here, and all one can really hope to provide, is training in intuition. One wants the student to get a feel for words, and the ways their meanings shift. I have never found a good way to do this. Socratic pursuit of definitions may help here, as well as some old Oxford-type ordinary language analysis. But how does one prevent this from sinking into sheer pedantry? I think sometime I may try to get the students to do cross-word puzzles, the kind with cryptic clues, for these puzzles make great play with ambiguities and linguistic tricks. But in any event, the point is to get the student to view his own use of language as an object—to view it critically. I have sometimes urged this on my students as 'second level thinking', but with only middling success. Much the same is true of the fallacies. The student should be able to see that they are fallacies by applying his ordinary tools of argument analysis. That analysis is all the theory he needs. It is a point of only incidental interest that some of these bad arguments belong to famous types, and have been given names. There is no particular point in learning a classification system for fallacies.

I have so far spoken of the components of the course from the point of view of their contribution to the grasp of theory or the enhancement of intuition. But there is also the matter of contagion and pleasant association, and I want to say a bit more about that, especially the pleasant association part. The great enemy here is tedium. I have said that studying reasoning is like cleaning up in preparation for a party. If the associations for the student are to be pleasant, then he must come to the party too. Which means, he must have the opportunity in the course to use his reasoning skills on interesting and important subject matter. This brings us to the problem of examples, for it is mainly through examples that tedium is kept at bay. Not enough attention, I think, is usually paid to this problem. The examples must illustrate the logical points, of course. But there are several other things they must do. They must be of independent interest to the student, and also importance, so that when he applies his logical tools to them he can feel that he is doing useful work. And on top of this, I think the examples should have, where possible, an independent educational value. This is partly because there are blank tablets out there in the classroom and this is an opportunity for us to write upon them, a chance which the true pedagogue cannot pass up. But also I think it contributes to the love of rationality if the student can see it functioning even handedly at all intellectual levels, and as opening up to him the chance to move to a higher one. So while we should not be above using examples from Ann Landers or silly letters to the editor, we should also be prepared to move higher, as high as the abilities of our students permit.

Perhaps this is enough to give an idea of the sort of course I have in mind. So I shall conclude by restating my main thesis. It is that in such a course the ability to reason well really can be taught.

Keep the faith!

Footnotes

[1]In this connection I confess to feeling rather sorry to see the term 'practical' creeping into discussion of critical reasoning courses and textbooks. People use the word for the wholly laudable purpose of distinguishing between teaching the ability actually to reason well and teaching the pure theory of correct reasoning. The course and subject is said to be practical because benefits will accrue in everyday life—the thing is useful. However, and this is the source of my concern, to refer to useful instruction in reasoning as

instruction in practical reasoning is to risk losing the term 'practical reason' in its Kantian sense of reason guiding conduct. But I shall not go into that on the present occasion.

ADVERTISING: ITS LOGIC, ETHICS AND ECONOMICS

Alex C. Michalos
University of Guelph

I. Introduction. The aim of this paper is to disclose some of the logical, ethical and economic features of contemporary advertising in North America. For reasons that are explained later, there appeared to be no satisfactory way to avoid ethics and economics, although the primary focus of this conference is logic.

After adopting a working definition of "advertising" in Section II, I show how the theory of public goods plus a few plausible assumptions would lead one to expect some deceptive advertising (Section III). Loto Canada advertising is considered as a case of deceptive advertising with a public sponsor (IV), and subliminal advertising is considered as a particular species of deceptive advertising (V). Finally, (VI) several criticisms of advertising and responses by two contemporary apologists are examined.

II. A working definition. According to the authors of a popular contemporary textbook on advertising, "Advertising is *mass communication* of *information* intended to *persuade* buyers so as to *maximize dollar profits*."[1] This definition has a number of implications that may not be immediately apparent or acceptable for all purposes. Although it is suitable for my purposes, some of its limitations should be mentioned before proceeding.

First, according to this definition advertisers are interested in communicating messages on a large scale. While we don't know how big "large" is, it may be assumed, for example, that at least not every display of products for sale will count as advertisement. Second, the immediate aim of the messages according to this definition is to get people to buy things. Thus, so-called social marketing or public service advertising is ignored. For example, Health and Welfare Canada's ads intended to get people to stop smoking and drinking are not covered by this definition. Third, the realization of the immediate aim of advertising is supposed to contribute directly to the final aim which is the maximization of dollar profits. This apparently presupposes that the individuals or firms sponsoring advertising have as their primary aim the maximization of dollar profits. I suppose many of them have that aim, but there is some evidence that neither maximization generally nor dollar profit maximization in particular are universally accepted goals.

By accepting the Littlefield and Kirkpatrick definition of advertising, I have limited my discussion in some ways and expanded it in others. The limitations have already been indicated. The expansion comes as a result of thinking about advertising in its North American setting as a socioeconomic institution. From this point of view, it is virtually impossible to untangle logical, ethical and economic issues in a logically tidy fashion. We just have to put up with some fudge.

III. From information to persuasion and deception.

Insofar as ads provide information more or less indiscriminately to great numbers of people, ads may be regarded as public goods. (To simplify things I usually use the single term "goods" as short for "goods and services." Strictly speaking ads seem to be typically more like services (e.g. information) than material goods.) There is a substantial body of literature on the theory of public goods. Such goods are supposed to be distinguishable from private goods on the basis of *either* of two characteristics, namely, jointness and non-exclusiveness.[2] To say that a good is characterized by jointness is roughly to say that using it does not imply using it up. To say that a good is characterized by non-exclusiveness is roughly to say that non-purchasers cannot be excluded. Information is a perfect example of a public good displaying the character of jointness, and clean air or national defense are examples of public goods displaying non-exclusiveness.

The fundamental problem concerning the provision of public goods is often referred to as the free-rider problem.[3] Since, without taking special measures, no one in a society can be excluded from public goods displaying non-exclusiveness, there is a temptation for (hypothetically self-serving) citizens to try to pass the costs of such goods on to everyone else. So, for example, Jones will leave his thermostat up because he will be able to enjoy the benefits of a national effort to conserve energy no matter what he does (One person's action has a negligible effect on the total picture.) and he won't have to bear the increased costs in chilly discomfort. Similarly, Smith will continue to throw his beer cans out the car window, to stay home on election day, and so on. If the environment is ever cleaned up or good politicians are ever elected, Smith will enjoy the benefits anyhow. Meanwhile, he lets the other people pay the tab. Smith and Jones, then, are free-riders.

If free-riders are regarded as the fundamental problem for the provision of public goods, coercion is usually regarded as the fundamental solution. People must finally be forced to pay taxes for public safety and fines for public pollution. For failing to be informed about the activities of their political leaders, they pay the price of polluted political processes. Of course in the best of all possible worlds, people would be aware of the nature of public goods, there would be no free-riders and no coercion. Part of the task of moralists is to help us get from this world to that other one—without, I would hope, leaving our corporeal bodies behind.

From the theory of public goods and the assumption that advertising involves the distribution of information, one may infer with some plausibility the reason for trying to make ads persuasive. If ad sponsors only distributed information then, unless their products really were superior to others *and* people could be counted upon to prefer superior products more often than inferior products, there would be no reason to suppose that the information would motivate people to buy their products. In fact, many products really are practically indistinguishable from others in their line, e.g. cigarettes, beer, bicycles, soap, and so on. (Neither producers nor consumers have unlimited discriminatory powers.) So, if ads were only informative, they would probably produce random purchasing of such products. People would be informed that, say, one soap is as good as any other. So there would be no incentive to shop carefully. What's more, there would be no incentive in the form of private profit for producers to advertise. Information about such products would be a clear public good and producers would all tend to take a free ride.

Apparently, then, from the point of view of producers reaping private profits, many ads must be persuasive. However, insofar as products are practically indistinguishable from others in their line, advertisers are left with the logically and morally outrageous task of designing ads to persuade people to differentiate indistinguishables and to prefer one to another! As Rosser Reeves put it:

> "Our problem is—a client comes into my office and throws two newly-minted half-dollars on my desk and says, 'Mine is the one on the left. You prove it's better.'"[4]

To avoid misunderstanding, let me emphasize that I am *not* claiming that it is good or smart to accept private enterprise, profit maximization, self-serving behaviour or even advertising itself.[5] What I have tried to do is show that certain assumptions about these things and the theory of public goods lead fairly directly to the conclusion that sometimes (not necessarily always) advertisers are committed to logically and morally bad practices. I have no doubt at all that this fact is probably more or less clear and more or less tolerable to different advertisers. As a class of human beings, I imagine they are no better or worse than the rest of us.

IV. Misleading Ads: A Case Study. The last section may give one the impression that deceptive advertising is only a product of private enterprise. So I would like to review a particular case of such advertising that involves the Canadian government. Although the advertising is sponsored by the government through a Crown Corporation, it is not public service advertising or social marketing.

The amendments to the Combines Investigation Act which became law in

December 1975 included several clauses concerning misleading advertising. The new clauses are supposed to "apply to all kinds of serious misrepresentation concerning products or services made to public, rather than merely to published advertisements. Not only the literal meaning of a representation, but also the general impression it conveys is to be taken into account."[6]

When I read the new provisions of the Act, I was impressed by the amount of protection the government was willing to give me. After failing to get the federal government to prosecute itself for patently misleading Loto Canada ads, I was impressed by the amount of deception the government was willing to practice. I can't go into all the details, but the following correspondence will demonstrate some of the logical and moral problems involved with misleading advertising.

The first letter was sent to "Box 99," which is the official complaint address in the Department of Consumer and Corporate Affairs.

> Box 99
> Dept. of Consumer &
> Corporate Affairs
> Ottawa, Ontario
>
> July 29, 1977
> Sir:
>
> I am writing to appeal to you to put a stop to the seriously misleading advertising of Loto Canada on television. The ads continue to be very attractive and undoubtedly persuasive, but they are clearly giving a distorted picture of reality.
>
> The ads make it seem as if the national lottery presented a good opportunity or chance to increase one's income with a windfall winning ticket. But since it is about 150 times more probable that one will die in an auto accident than that one will hold a winning ticket, the chances of the latter should not be described as good. If they are good, then the others are so much better that it is foolish to buy a ticket. It would seem unlikely that one will be alive long enough to collect it. That of course is false. We are much more secure in our autos than this scenario would suggest. The truth is that it's highly unlikely that any given individual will hold a winning ticket. So it is immoral and should be illegal for our government to create a quite different impression day after day on national television.
>
> The credibility of government information releases is always under some strain in virtually all societies. But with blatantly misleading advertising the tension is needlessly increased. The government cannot expect to be able to con us with phoney ads at one moment and mobilize our support for national unity, belt tightening or other serious problems the next.
>
> At the very least I urge you to see that the odds of winning are always

in plain sight wherever lotteries are advertised. Honesty in advertising must be taken seriously.

Sincerely,

Alex C. Michalos.

ACM/sdm
cc: Pierre Trudeau
    A. Alan Borovoy

I was notified that my letter had been forwarded to another office, and I sent the next letter to that office.

Chief of Operations
Marketing Practices Branch
Place du Portage, Phase I
68 Victoria Street
Hull, Quebec

August 30, 1977
Sir:

Concerning my complaint about Loto Canada TV ads which was forwarded to you office (File No. TP 100.402), I call your attention to the paragraph below from the Combines Investigation Act. I believe the most frequently broadcasted TV ads for Loto Canada violate this section of the Act by not explicitly stating the chances of winning.

Surely the intention of Parliament in this section is to preclude misleading advertising that stimulates people to act in the absence of full knowledge of the likely consequences of their action. The point of the TV ads is precisely to get people to do what Parliament was trying to prevent them from doing, namely, buy on impulse rather than on the basis of a rational calculation of the likely benefits of the purchase. Accordingly, I urge you to do your duty and see that these illegal and immoral ads are stopped.

Combines Investigation Act

Section 37.2(1): No person shall, for the purpose of promoting, directly or indirectly, the sale of a product, or for the purpose of promoting, directly or indirectly, any business interest, conduct any contest, lottery, game of chance or skill, or mixed chance and skill, or otherwise dispose of any product or other benefit by any mode of chance, skill or mixed chance and skill whatever unless

    (a) there is adequate and fair disclosure of the number and value of the prizes and the chances of winning in any area to which prizes have been allocated.

       Sincerely yours,

       Alex C. Michalos

ACM/sdm

I received the following response.

  Director of Investigation
  & Research
  Combines Investigation Act
  Ottawa-Hull
  K1A OC9

  October 24, 1977
  Dear Mr. Michalos:

Thank you for your letter of October 19, 1977, concerning your earlier complaint against Loto Canada Television advertising.

Section 37.2(1)(a), of the Combines Investigation Act, states that in any contest promoting directly or indirectly the sale of a product there must be the following:

> "adequate and fair disclosure of the number and approximate value of the prizes and of the area or areas to which they relate and of any fact within the knowledge of the advertiser that affects materially the chances of winning."

With respect to this section of the Act, we have reviewed past and present Loto Canada advertising and we believe that all the requirements of the section have been met. The facts which would materially affect one's chances of winning, i.e. the number of tickets available to be sold and the number and value of the prizes, were disclosed. It was also the Director's opinion that adequate and fair disclosure occurred when the above information was made freely available to the public in newspapers and point of purchase display material during the run of the contest.

Should you have any additional questions on this matter please do not hesitate to contact this office.

       Yours very truly,

       Douglas G. Fraser
       Marketing Practices Branch

DGF/kc

To that I replied:

> Douglas G. Fraser
> Marketing Practices Branch
> Consumer and Corporate Affairs
> Ottawa, Ontario
>
> 77 10 31
>
> Dear Mr. Fraser:
>
> I am very disappointed by your conclusion regarding Loto Canada advertisements. You apparently believe that if 99 percent of an advertising campaign is misleading but one percent is not, then the campaign is fair. This is outrageous.
>
> Not once has Loto Canada advertised the odds of winning to a national TV audience. Occasionally a TV news reporter will mention the problem. You don't even perceive it as a problem. Posters can be found on most government buildings urging people to buy, but the posters never give the odds. So in most display areas, the odds are not "freely available" as you say. People have to go out of their way to find the odds of winning, but they are bombarded with advertising material urging them to "buy a ticket on their dream."
>
> The advertising is not fair, and you ought to be ashamed of yourself if you are not able to perceive its serious bias and unfairness. We are being systematically misled and encouraged to buy on impulse by Loto Canada and Wintario ads, and it is your responsibility to prevent such things. But you won't. What a sad state of affairs. What a pathetic way to carry out an oath of office or run a government.
>
> Sincerely yours,
>
> Alex C. Michalos

There was no reply to this last letter. I have since written to some MPs and received some sympathetic replies, but there has been no action by the government. If we assume that, for example, Mr. Fraser and others in the Marketing Practices Branch are just ordinary honest civil servants, then it must be granted that they see nothing in Loto Canada advertising that violates the Combines Investigation Act, Sec. 37.2(1)(a). At a minimum that tells us that the determination of misleading advertising is by no means a straightforward issue. I have already suggested what it tells us at a maximum in my last letter.

Since the above letters were written, a battle has raged between Provincial and Federal governments over the right or wisdom of Federal versus Provincial lotteries. Both levels of government see the lotteries as good sources of revenue, and both are apparently going to fight to keep the money coming in. There is no noticeable difference in the advertising for the two levels of government.

Given the wide variety of ways to win various sums of money, I now doubt that it will help much to give the odds of winning each sum. What is required is a clearly visible report of the *expected value* of every ticket purchased. If, for example, buyers knew that any ticket had an expected value of 50 cents or whatever and cost a dollar, then they could make an informed choice. Maybe most buyers would be willing to pay for the fun of the gamble. I certainly have no objection to people spending their own money in that way. But at present it's practically impossible to make an informed choice about a lottery ticket's value, and that is intolerable.

V. Subliminal Advertising. In the third section it was claimed that people pursuing private profit would often be engaged in deceptive advertising, and in the previous section it was claimed that government agencies also engage in such practices, wittingly or not. In this section I want to address the problem of subliminal advertising. This is the sort of advertising that Key claimed involved "intuitive or insight logic."[7] It is also the sort Johnson and Blair seem to have been thinking of when they wrote that

> ... although advertising is an attempt to persuade, the type of persuasion generally used is not *rational*. Instead, advertising attempts to persuade us by appealing to our emotions (our hopes, fears, dreams), to the vulnerable spots in our egos (our desire for status and recognition), by applying pressure to the tender areas of our psyches.... In sum, *advertising has a logic of its own*. Thus, learning how to evaluate ads from the standard logical point of view becomes a gratuitous exercise.[8]

Key's *Subliminal Seduction*, like Vance Packard's *The Hidden Persuaders*[9] fifteen years earlier, stimulated a lot of discussion about advertising tactics. Key refers to subliminal perception as any "sensory inputs into the human nervous system that circumvent or are repressed from conscious awareness."[10] The most famous experimental proof of such perception involves the flashing of brief messages on a screen with a tachistoscope.[11] Although

> Once we accept my basic premise and that is that advertisers are entitled to do more than simply deliver a factual description of their product and service and price; once one acknowledges that *an advertiser is entitled* to appeal for the viewer's demand, *to create demand* for a class of goods and for his goods in particular, once we acknowledge that as being a principle—then *we have got to allow the advertiser to appeal to the conscious and sub-conscious appetites of the viewer.*[17] (Emphasis added.)

Insofar as the logic of advertising is the logic of subliminal perception and "subconscious appetites," I'm inclined to regard the subject as more suitable to empirical investigation than to conceptual analysis. Unfortunately, I just don't know how far the logic of advertising is a matter of such perception and appetites.

It was suggested (during the discussion of this paper at the Conference) that subliminal advertising involves a unique sort of inference or implication relation. Maybe so. But I think what's involved is more a matter of interpretation than inference. Once a particular interpretation has been made of a feature of an ad, the move from that interpretation to the conclusion (practical or cognitive) planned by the advertiser may be a move that's indistinguishable from ordinary inference. The history of attempts to clarify concepts of implication, inference, entailment and so on is such that I am reluctant to wade into that sea of troubles unless absolutely forced.

At this point I want to set out in a slightly different direction, and to consider in detail a set of alleged criticism of advertising and responses offered to the critics by Littlefield and Kirkpatrick (hereafter LK).

VI. Criticism, replies and comments. Before I begin the series of arguments in this section, it may be worthwhile to expand a point that was just barely suggested in the second section. People have traditionally defended free speech or expression on two grounds. For some it has been a matter of moral principle that people ought to be allowed to express their views without fear of reprisals. While it may be a defeasible moral right, it is nevertheless a fundamental moral right, like the right to life, for example. For others, free expression has been defended as a matter of epistemological good sense. These people are more interested in the pursuit of truth and the avoidance of falsehood than they are in moral principles, and they see the free expression of ideas as a necessary condition of their epistemic aims.[18]

Of course, no one has to choose between these two different grounds for defending free expression, but historically I think people have tended to lean toward one or the other as especially weighty. (If one is primarily interested in the free expression of fictional or visual material (stories, films, paintings, sculptures) then one's defence might run more smoothly from moral grounds, while if one is primarily interested in the free expression of non-fictional or descriptively accurate material then one's defence might run more smoothly from epistemic grounds.) The point I want to emphasize here, however, is that the two grounds often coalesce. In particular, objections to advertising practices may involve epistemological (or narrowly logical) and moral principles at the same time. Indeed, I suspect that this is typically the case. Hopefully, this will become clearer as the discussion proceeds. Let us turn immediately to a consideration of LK's critique.

1. LK begin by answering the charge that advertising is often "false, deceptive, and misleading, and that it conceals information which should be revealed and omits limitations and comparative disadvantages of the item advertised." In their view, "There is no justification for false, deceptive, or misleading advertising."[19] They don't deny the charge at all, and they claim that self-regulation and enlightened self-interest (buyers must want to return) tend to minimize such practices. In a very revealing passage they tell us that

> To tell advertisers to limit themselves to non-emotional, non-persuasive advertising would be to take a step in a direction repugnant to most of us. "It is not the primary function of advertising to educate or

to develop reasoning powers."[20]

Comment. We have already seen how an advertiser might be led down the garden path to deception. The question is: Are there any good reasons for thinking that there is anything "repugnant" about insisting on "non-persuasive advertising"? Why should LK think that is demanding too much? Presumably they have given their answer to these questions a few pages later. "Persuasion and influence here," they write, "are just as ethical as in politics, religion, or education."[21] That is, they believe that there is nothing in principle morally wrong with trying to be persuasive as well as informative.

One would like, I suppose, to respond that there is something better about trying to persuade people to vote, worship God or get an education. But by the time the words are uttered or written, I begin to have second thoughts. *A priori* I doubt that any old persuasive case made in behalf of any old political, religious or educational cause must be somehow morally superior to any old case made for any old product or service. There are too many worthless and even dangerous political, religious and educational causes and too many worthwhile marketed products and services to permit full-scale whitewashes. So I think we have to agree that there is nothing in principle wrong with persuasive advertising, and objectionable cases will have to be tracked down and eliminated one at a time.

2. In response to the charge that "advertising confuses and bewilders more than it helps," LK claim that "differences of opinion are a basic element in our mores and in our norms." Besides, they don't believe anyone can be "objective about his brand any more than can . . . a bridegroom about his bride . . ."[22]

Comment. The latter claim would be self-defeating if it were true, because the claim itself would lack objective persuasiveness. But it's plainly false. Everyone has all sorts of "objective" information about his or her most cherished persons or things. For example, I know the colour of my wife's hair and eyes, her height and weight, how she prefers her tea, and so on. Loving someone or something is not the same as being struck dumb. Even the most ardent fans are often prepared to admit that their team doesn't have a hope in hell of winning, and there would be no sense at all in anyone's favouring the underdog unless there were a more or less objective assessment of just who *is* the underdog.

Some years ago David Braybrooke leveled the charge of confusion against corporations in an excellent article called "Skepticism of Wants, and Certain Subversive Effects of Corporations on American Values."[23] He mentioned in particular "the systematic abuse of sexual interests, so that people have their wants for automobiles and all sorts of other things seriously mixed up with their sexual desires."[24] On top of that he claimed that

> . . . corporations not only assist in confusing the public about what it might want, they also obstruct institutional remedies for the lack of information that leads . . . consumers into misjudgments about

wants.

> ... How shameful to find, besides the automobile companies dragging their feet about safety standards, the tire companies doing the same thing; the grocers and packagers objecting to truth-in-packaging; the credit firms protesting against truth-in-lending.[25]

Braybrooke's primary concern was as much epistemological as moral. Allegedly incorrigible first-person reports about wants, he argued, could be muddled and in need of revision given the heavy hand of corporate advertising. What's more, it seemed to him (as it does to me) to be morally wrong for corporations to "obstruct institutional remedies" in the ways he mentioned. It is one thing to have differences of opinion, but something else to prevent the unbiased assessment of claims and counter-claims.

3. In response to the charge that advertising is often "vulgar, and in poor taste," LK claim that "advertising has no responsibility to raise consumers' tastes, to preach, to try to elevate." In fact, they insist that a wise advertiser "should determine what your tastes are" and "he should then cater to those tastes ..."[26]

Comment. I wonder first, just whose responsibility LK suppose it is "to raise consumers' tastes" and "to try to elevate." I suppose they would want to claim that it is the business of teachers, professors, theologians and moralists "to try to elevate." Advertisers are in a different business, the business of selling products for profit. Therefore, they should have nothing to do with elevating people—unless it's a matter of elevated shoes, airplanes, and so on.

This is a familiar piece of buck-passing that must be met head-on. It is a mistake to think that things like values, norms and morality must exist in some proper ontological pigeon hole of the universe which one can dip into or avoid pretty much as one pleases. There is no good or evil in the abstract. Good and evil, values if you like, must be attached to things, actions, people and so on if they are to have any existence at all. Thus, for example, if there is any moral behaviour then it will be found by looking at ordinary behaviour from a moral point of view. If an advertiser produces ads in which false claims are intentionally made then the advertiser is a liar. All and only people who intentionally make false claims *can* be liars. They will be lying advertisers, lawyers, philosophers, plumbers, housewives or whatever, but they will be liars all the same. Therefore, and this is the main point, in order to have a world in which there are no liars, advertisers must stop lying when they are practicing their trade, lawyers must stop lying in their work, housewives must stop lying, and so on. There is no other way to make a world without liars.

The mistake involved here seems to be in regarding moral behaviour as a special kind of sociological role playing. However, being a morally decent person is not analogous to being a butcher, dentist or school teacher. It is not another role or alternative hat one slips on now and then. Insofar as analogies help, one may say that being a morally decent person is like being clean in the literal sense of well-scrubbed. There is no once-of-a-life-time bath one can take that will keep one clean forever. Every day brings new dirt. However, when one is clean or dirty, one is clean or dirty at dinner, selling shoes or buying hamburger. Being clean or dirty is not a sociological role in addition to the consumer's role, the farmer's role and so on. It is an aspect or feature of anyone operating in any of those roles. Just so, being a morally decent person or a person of high moral character is an aspect or feature of a person no matter what his or her sociological role. And it is an aspect that must be forever cultivated.

Insofar as one believes, for instance, that a world without liars is preferable to a world with liars, one ought to recognize one's responsibility for bringing about such a world. People who perform morally good actions are performing public services *par excellence* (which is not to say that agents receive no private benefits from such actions). Whenever one resists the temptation to lie, for example, one is engaged (in a limited way to be sure) in building a better world. It is the business of advertisers, bakers and all people in any role whatever "to try to elevate" the world by adopting a moral point of view *in that role*. To say that they might adopt a moral point of view when they are in some other role, like children in Sunday school, is to say that they don't know what it means to adopt a moral point of view or to try to be a morally decent person.

LK's second claim, namely, that advertisers should cater to consumer's tastes, must be understood conditionally in order to avoid contradicting their first commitment to persuade people to buy products so they can make a profit. Their aim must be to use people's tastes as instruments for manipulating people's consumption habits. Nothing in their position suggests that they would not mould people's tastes to suit their own purposes if they thought they could get away with it. Indeed, just the opposite is true. They are committed to persuading people to buy their products. In LK's own words: "In a sense, demand must be stimulated continuously." "The advertiser's hope is to make prospects dissatisfied with their present status and to keep current customers satisfied."[27] Insofar as anyone has a taste for something that is incompatible with an advertiser's product, the latter must try to alter the taste, the product or the appearance of the product. Since his primary objective is profit maximization, any of these three alternatives would seem to be live options.

4. Advertisers have been charged with getting "consumers to buy what they (a) do not need, (b) should not have, and (c) cannot afford."[28] But LK reply that nothing is ever bought that is "not in response to an admission of *need*." Besides,

> ... who knows what Mrs. Homemaker needs and can afford better than Mrs. H. herself? No one, of course. Just try to get her to buy something that will not (a) protect or (b) enhance her self concept."[29]

Comment. This is an incredible passage, but a fair reflection of LK's position. Roughly speaking, they have only substituted "need" for "want" in the old cliche "We only give the public what it wants."[30] The latter claim was thoroughly discredited by Braybrooke in the article cited earlier. But what can we make of the suggestion that sellers only give buyers what they need? Does anyone need Hostess Twinkies, Pringles, fat ties, thin ties, short skirts, long skirts, and so on? People have a need for food in order to live; but for Pringles? For grapefruits with skins that belong on footballs? That can't be true.

LK's claim about the role of the protection and enhancement of one's self concept in marketing is probably not as outrageous as it may appear. Basically their view seems to be only that people will not pay money to be assaulted in any serious or threatening way. That's weaker than what seems to be claimed in the quotation above, namely, that people will only pay money for things that protect or enhance their self concept. I doubt that they imagine that, for example, everyone buying bananas, bandaids and buttons is somehow building up his or her self concept.

5. In response to the charge that advertising helps create a society of "greedy, self-centered individuals who worship materialism," LK reply that "The great majority of U.S. consumers believes that each person should expand his needs and then gratify them." What's more, however, they insist that "the purpose and responsibility of advertising are to make ultimate consumers want to consumer more."[31]

Comment. The sentence about "the great majority of U.S. consumers" leads me to suspect that LK have a peculiar notion of "needs." They seem to be claiming that one, anyone and everyone, ought to have more needs. For example, I suppose, they would want to say that I ought to need a Cadillac, hair dryer and over-the-calf socks. (In the latter case I would also need bigger calves—or garters.) On the contrary, I can't think of any good reason for having an obligation to need such things. I even suspect that the idea of obligations to need things is incoherent. So, *a fortiori* I think the claim that a majority of Americans believes I have such obligations is completely unfounded and farfetched.

The second quotation from LK seems to grant the charge to which they are replying. The responsibility of advertisers, as LK see it, is to make people "want to consumer more." It is not claimed, you may notice, that advertisers should try to make us want to consume more *if* that suits our tastes, *if* that's what we want or *if* that's what we need. The obligation is categorical. The name of their game is "Make people want to consume more." They explicitly claim that "Materialism should not be an end—it should be a means to even better ends."[32] But it's not clear what "better ends" LK might have in mind. Whatever we have, their aim is to make us want more. For advertisers with this view, the best of all possible worlds is one in which all human problems and solutions are manufactured and sold in the marketplace. It is a world in which everyone believes that he or she has some problem that can be solved by buying something that someone else wants to sell. It is a merchandiser's paradise. Indeed, LK suggest that the dream is not too far away.

Every individual who wants to can be just as individualistic as he or she prefers. And there are enough dollars and enough different goods and services in our affluent society to afford wide ranges of choice.... Where else, indeed, in the world can the consumer find the assortment of merchandise and services with which to express his individuality?"[33]

How easily they neglect the poor slobs who might want to "express their individuality" without buying something. The very idea of such individuals seems to have escaped these authors completely.

6. It has been charged that advertising constitutes a severe constraint on the content of media which rely on its revenue to stay in business. For example, because advertisers use TV programs as means of getting people to sit still for their ads, controversial programs or programs revealing views about the world that are incompatible with ads are systematically eliminated. LK reply that "commercial medial can be 'free' of government subsidies, 'free' of political control because of dollars from advertisers.... Prices of media would have to be higher if there were no advertising."[34]

Comment. Apparently LK grant the charge but believe that constraints by advertisers are less objectionable than constraints by government, and there must be some constraints. Since some Canadian media are not free of government subsidies *or* advertisers, we may have the worst of both worlds. The mind boggles at the prospect of having all media run on the model of Pravda, but one can hardly be sanguine about the continous parade of reminders of yellow teeth, bad breath, smelly armpits, flakey hair, irregular bowel movements, and so on. It seems to me, however, that this is a false dichotomy. CBC radio has no ads but is not constrained by the Canadian government any more than, say, CBS is constrained by the American government. The Trudeau administration has threatened the CBC through the CRTC and otherwise, but the Nixon administration was at least as difficult for CBS. It is also possible to sustain media outlets with private subscriptions, as we do with some journals, radio and TV stations. Finally, one can always withhold one's support (e.g. change the channel, avoid the product) or take action against offensive outlets (e.g. join citizen action groups opposed to misleading advertising, obscene displays, and so on). Granted that there must be constraints, one doesn't have to be on the receiving end all the time.

7. In response to the charge that ads stress *"insignificant* product details, *minor* product differences, *unimportant* product changes," in a word, trivia, LK claim that what's minor today may be major tomorrow, that what's minor to you may be major to someone else and that what's perceived as major *becomes* major with increased consumption.[35]

Comment. Since ads are often intended to perform the logically impossible task that I earlier described as differentiating indistinguishables, the present charge would seem to be practically a truism. When products are essentially the same, only trivial differences will be discoverable. The point of LK's reply seems to be that if, for example, people are willing to pay two or three times as much for Bayer aspirin as they are for aspirin *simpliciter*, then it is at least misleading to regard the brand name as a trivial feature of the product. But that seems to be irrelevant to the critic's point. The latter seems to be that from the point of view of the effectiveness of the product or that for the sake of which the product is purchased, the brand name is an unimportant feature. (To simplify matters I am ignoring the fact that many drugs are purported to have a 30 percent placebo effect and that for some people the effectiveness of Bayer aspirin may be greater than the effectiveness of other brands.) Hence, by emphasizing the brand name, advertisers are guilty of trying to make something (significant) out of nothing (significant). Again, that is objectionable on epistemological and moral grounds.

8. It has been charged that advertising wastefully increases the cost of products. But LK claim that advertising represents "the shortest way to the market," to a mass market at least. It would be far more expensive to try to reach the same number of people with personal selling, house by house, person by person. They also insist that in theory effective advertising leads to increased sales volume which leads to lower per unit costs and the possibility of lower prices.[36] They grant, however, that "there is waste in advertising just as there is waste in competition," and then they wax poetic.

> If advertising were outlawed, something would take its place, and that something would most probably be more wasteful and more expensive. Actually, attacks on *advertising* are really attacks on our system and structure of *business*. Advertising is a part of and in harmony with our free enterprise system. Our free enterprise or competitive system is the cause of advertising, not the result.
>
> Abolish advertising because it is wasteful and competitive? Then abolish competition.[37]

Comment. As a former Fuller Brush man, I'm prepared to accept the claim that almost anything is more efficient than door to door selling. But personal problems aside, I would accept LK's first claim. Their theoretical defence of advertising is theoretically unexceptional, but not very useful in fact. What we would like to know, but don't, is the relative frequency with which the option of lowering prices is adopted over the option of reinvesting the new profits, distributing them to stockholders, employees, and so on. It would also be useful to know how wasteful advertising practices are, not necessarily in relation to non-advertising activities but in relation to some hypothetical optimum. (Presumably there is a vast literature on the return-on-the-dollar of various sorts of advertising, although I'm unfamiliar with most of it).

The most interesting part of LK's remarks in response to the charge of wastefulness is their claim, hardly necessary in this context, that our "competitive system is the cause of advertising." Does competition for market shares entail advertising? Could there be a competitive marketing system without ads? From a logical point of view, of course the two ideas are separable. We have already contrasted door to door marketing with marketing through advertising. In fact many producers simply produce their products and make them available to purchasers without advertising as that term has been defined here, e.g. farmers in community markets or roadside stands. Unless one loosens up the definition of advertising to include any sort of display of products for sale, it should be possible to easily multiply such examples of non-advertising marketing.

It is illegal to advertise some services in some areas, although the services themselves are quite legal, e.g. legal and medical services. While competition is probably far from the minds of many professionals, a spokesman for the Canadian Medical Association once said in a radio interview that one must remember that doctors are small businessmen. (Some of them are relatively big businessmen.)

Granted that competitive marketing does not logically imply advertising, the former seems to be a major contributing factor toward the existence of the latter. After all, it doesn't require much imagination to realize that, for example, if farmer Brown can sell his corn by merely putting it on display in a roadside stand, then he can increase the chances of a sale by increasing the visibility of his stand with big signs, by placing some signs far enough away so people can prepare to stop and finally by putting "signs" (ads) in news media to get people to make a special trip out to the stand. If they like his corn, maybe they will go for his chickens too. Maybe people that won't go for the corn will go for the chickens. Given the aim of selling something for profit in a world of scarce resources, it's difficult to imagine anything arising more naturally than advertising. That may be what led LK to see advertising and competition inexorably connected.

It is perhaps worthwhile to add here that I suspect a better case can be made for allowing advertising on the basis of a Principle of Liberty than on the basis of competition. By a "Principle of Liberty" I mean something like the maxim that people ought to be allowed to do whatever they wish as long as it doesn't harm anyone else. Not many people are likely to object to that idea. It follows immediately, then, that insofar as advertising is harmless, it is allowable, and much of it probably is harmless. Perhaps this is the sort of argument LK had in mind when they mentioned "free enterprise" in conjunction with competition. Nevertheless, neither "free enterprise" nor a Principle of Liberty implies advertising.

9. Several charges have been leveled against advertising as a monopolistic force in the marketplace. For example, advertising has been charged with creating barriers to entry for new products or firms, discouraging price competition and contributing to large-scale economic concentration.[38] LK's general position with respect to such criticisms is that monopolies came before advertising. While they don't and shouldn't deny that advertising represents some sort of a barrier to entry, they mention several others that may be at least as significant, e.g. "inadequate capital; lack of a full line of products; lack of competence, either manufacturing *or* marketing; channel difficulties, such as unavailability of essential distributors, or the magnitude of the job of building a dealer organization; patents."³⁹

Comment. LK's response is perilously close to claiming that advertising is *not* objectionable because other things *are* objectionable. Whether or not that is their argument, it is obviously unsound. *A priori* I suppose that anything that gives one a marketing advantage over competitors might contribute toward monopoly, perfect competition or something in between, depending on the total distribution of advantages and disadvantages. So it's misleading to claim that any particular marketing advantage, as for example a good advertising scheme, is *on its own* a monopolistic force. In principle such schemes could bring about perfect competition if all other competitors were lucky enough to create equally advantageous schemes. Of course in fact some producers, for one reason or another, can mount more successful advertising campaigns than other producers, and the latter have nothing to compensate for their weakness. Some people are economically wiped out in such cases, even though the losers may be more efficient operators than the winners on a dollar for dollar or product for product basis. That's the sort of thing most small businesspeople, which means most businesspeople, want to see prevented; but it happens. Legislation like the Combines Investigation Act provides some protection against big or unscrupulous operators, but, as suggested earlier in the case of Loto Canada ads, the legislation is not self-implementing.

VII. Conclusion. The aim of this investigation was to disclose some of the logical, ethical and economic features of contemporary advertising in North America. Given such a diffuse goal, it was fairly easy to hit the mark. After adopting a working definition of "advertising," I showed how the theory of public goods plus a few plausible factual assumptions would lead one to expect some deceptive advertising. Loto Canada advertising was reviewed as a case of deceptive advertising with a public sponsor. Subliminal advertising was briefly reviewed as a particular sort of deceptive advertising. Finally, several criticisms of advertising and responses by Littlefield and Kirkpatrick were examined.⁴⁰

Footnotes

¹J. E. Littlefield and C. A. Kirkpatrick, *Advertising: Mass Communication in Marketing,* Boston: Houghton Mifflin Co., 1970, p. 100.

²M. Olson, "The Plan and Purpose of a Social Report," The Public Interest, 1969, p. 94. See also M. Olson, *The Logic of Collective Action,* Cambridge: Harvard University Press, 1965.

³R. N. McKean, "Collective Choice," *Social Responsibility and the Business Predicament,* ed. J. W. McKie, Washington: The Brookings Institution, 1974, pp. 109-134.

⁴Quoted from R. H. Johnson and J. A. Blair, *Logical Self-Defense,* Toronto: McGraw-Hill Ryerson, 1977, p. 222.

[5]I have argued against maximization policies and self-serving in *Foundations of Decision-Making*, Ottawa: Association for Publishing in Philosophy, 1978.

[6]Canada, Consumer and Corporate Affairs, *Proposals for a New Competition Policy for Canada: First Stage*, Ottawa, 1973, p. 5. Excellent discussions of misleading advertising and the Combines Investigation Act may be found in D. N. Thompson, "The Canadian Approach to Misleading Advertising," *Problems in Canadian Marketing*, ed., D. N. Thompson, Chicago: American Marketing Association, 1977, pp. 157-184, and W. T. Stanbury, *Business Interests and the Reform of Canadian Competition Policy, 1971-1975*, Toronto: Methuen Pub., 1977.

[7]W. B. Key, *Subliminal Seduction*, New York: The New American Library, 1973, p. 11.

[8]Johnson and Blair, op. cit., p. 218.

[9]V. Packard, *The Hidden Persuaders*, New York: David McKay Co., 1957.

[10]Key, op. cit., p. 18.

[11]Key, op. cit., p. 21, and S. J. Arnold, J. G. Barnes and K. B. Wong, *Brief to the Canadian Radio-Television Commission on Subliminal Perception: Implications for Regulation*, dittoed, March 11, 1975, p. 11.

[12]Key, op. cit., pp. 33-34.

[13]Canadian Radio-Television Commission, Research Branch, *Subliminal Perception and Subliminal Advertising: An Overview*, dittoed, March 1975, p. 29.

[14]Canadian Radio-Television Commission, *Hearings on Proposed Amendments to the Television Broadcasting Regulations (Advertising "Subliminal Technique")*, dittoed, March 11, 1975, pp. 446-447.

[15]From a letter of the ACA Counsel, C. R. Thomson, to the CRTC, February 17, 1975, p. 1.

[16]From a letter of the President of CAAB, R. E. Oliver, to the CRTC, February 24, 1975, p. 2.

[17]C. R. Thomson in the CRTC *Hearings*, pp. 458-459.

[18]For an epistemic approach see J. S. Mill, *On Liberty*, London: Parker, 1859, and for a strictly moral approach see the United Nations Universal Declaration of Human Rights or the Canadian Bill of Rights in P. E. Trudeau, *A Canadian Charter of Human Rights*, Ottawa: Information Canada, 1968.

[19]Littlefield and Kirkpatrick, op. cit., p. 115.

[20]Ibid. The quotation is from S. V. Smith, "Advertising in Perspective," in J. W. Towle, ed., *Ethics and Standards in American Business*, Boston: Houghton Mifflin Co., 1964, p. 174. For a good analysis of self-regulation in Canada, see M. S. Moyer and J. C. Banks, "Industry Self-Regulation: Some Lessons from the Canadian Advertising Industry," *Problems in Canadian*

Marketing, ed., D. N. Thompson, Chicago: American Marketing Association, 1977, pp. 185-202.

[21]Littlefield and Kirkpatrick, op. cit., p. 115.

[22]Ibid.

[23]D. Braybrooke, "Skepticism of Wants, and Certain Subversive Effects of Corporations on American Values," **Human Values and Economic Policy,** ed. S. Hook, New York: New York University Press, 1967, pp. 224-239.

[24]Ibid., p. 230.

[25]Ibid., pp. 230-231.

[26]Littlefield and Kirkpatrick, op. cit., pp. 116-117.

[27]Ibid., p. 124 and p. 102.

[28]Ibid., p. 117.

[29]Ibid.

[30]On the differences between needs and wants see Michalos, op. cit.

[31]Littlefield and Kirkpatrick, op. cit., p. 117.

[32]Ibid.

[33]Ibid., p. 118

[34]Ibid.

[35]Ibid., pp. 118-119.

[36]Ibid., p. 119.

[37]Ibid., pp. 123-124.

[38]Ibid., pp. 121-123.

[39]Ibid., p. 122.

[40]I would like to thank Rodrigue Chiasson, Acting Director-General of the Research Branch of CRTC for providing a copy of the Examination File on "Regulation and Policies Proposed Amendment to the Television Broadcasting Regulations, Ottawa Hearing, March 11, 1975.

EVALUATION OF INFORMAL LOGIC COMPETENCE[1]

Thomas N. Tomko and Robert H. Ennis
University of Illinois, Urbana-Campaign

One of the distinguishing features of the developing informal logic movement is its concern with pedagogy. Informal logic teachers want to know what skills and concepts are important for their students to know, how these skills and concepts can best be taught, and how one can determine if they have been successfully taught. Certainly the answers to the latter two questions will require some empirical research. In this paper we will examine tests and evaluation procedures in the hope that we can shed some light on these two questions. Our intended audience is people who have the above concerns, but who are unfamiliar with available tests and/or with the field of testing.

TESTS

One naturally turns to tests as a starting point for the evaluation of teaching. Although not all people agree on the value of tests, they are a very practical and widely-used evaluation tool.

Currently available tests

The Watson-Glaser Critical Thinking Appraisal.
This test is perhaps the most widely-used instrument in the area of logic and critical thinking. It has two forms, Ym and Zm (revised in 1964), each 100 items long and divided into five subtests:

> Inference—ability to discriminate among degrees of truth, falsity, or probability of inference drawn from given facts or data

Recognition of Assumptions—ability to recognize unstated assumptions in given assertions or propositions

Deduction—ability to reason from given premises, recognition of logical implication

Interpretation—ability to weigh evidence and to discriminate among degrees of probable inference

Evaluation of Arguments—ability to discriminate between strong and weak, important and irrelevant arguments

The type of answer to be selected varies with the sub-tests. For example in Test 1 (Inference) students must decide, after reading a paragraph, whether some proposed inferences are true, probably true, probably false, false, or cannot be judged due to insufficient data. In Test 2 (Recognition of Assumptions), the examinee must decide, given a statement, whether or not another statement is "presupposed or taken for granted." For example, given the statement "Let us immediately build superior armed force and thus keep peace and prosperity," one must decide whether the proposed assumption "The building of superior armed force guarantees the maintenance of peace and prosperity" is "necessarily" made or not made. In Test 3 (Deduction) and Test 4 (Interpretation), one must decide whether a conclusion does or does not follow. In Test 5 (Evaluation of Arguments), one must decide whether short, two- or three-sentence arguments are strong or weak. The questions in each test are preceded by sample items and an explanation of what the student is to do.

The test does not cover some aspects of critical thinking which one might wish to cover. For example, no semantical skills or concepts are covered, so there are no questions dealing with definition or ambiguity. There are no questions dealing with the reliability of observation statements or statements made by authorities.

To the extent that the Watson-Glaser test measures the ability to deal with induction, it has problems that also trouble the Cornell Critical Thinking Tests (CCTT), discussed below. In order to answer some of the items correctly, examinees must have certain background knowledge. If they do not, they may answer incorrectly even though the *way* they reached the answer might be judged to be acceptable. This problem appears in just about any induction test. We have found that it is very difficult to construct test items on inductive reasoning that have one answer that is clearly the best. One can almost always defend alternative answers by making certain reasonable assumptions that were not foreseen by the test maker.

Another problem with the Watson-Glaser test is its concept of an *assumption*. The directions to the section, "Recognition of Assumptions," say, "If you think the assumption is *not* necessarily taken for granted in the statement, make a heavy line under 'ASSUMPTION NOT MADE' on the Answer Sheet." (p. 4) We can probably neglect the problem of referring to the *proposed* assumption as an "assumption," but there is a more serious problem. The wording encourages the test taker to look for something that is *necessarily* taken for granted. Possibly presuppositions (of the type discussed by Strawson, 1952) are "necessarily" taken for granted, but premise-type assumptions are not. There is always another and different premise or premises to fill a real gap in an argument. So no premise-type assumption is necessarily taken for granted (Ennis, 1961, elaborates this point).

A third problem is that one's answers to the items in Test 5, Evaluation of Arguments, often depend on one's politics and values. It does not seem fair to mark a person wrong because of the person's politics and values. For example, Items 89 and 91 are keyed weak arguments on the issue of whether the United States government should try to inform the public of sought-after results of its scientific research programs in the areas of "new weapons, equipment, devices, etc." These items are:
    89. No; some people become critical of the government when widely publicized projects turn out unsuccessfully.
    91. Yes; the projects are supported by taxes and the general public would like to know how their money is to be spent.

We agree with the key on Item 89 and disagree on Item 91, but we can see how people with a value position different from ours might well answer just the opposite. We are not here urging a relativistic ethics, but do not think that a person's critical-thinking score should depend on that person's politics and values in such cases.

A fourth problem for some is the orientation of the test to the United States, exemplified in the above-mentioned issue in Test 5. Students from other English-speaking countries might be penalized for being less familiar with United States government structure, policies, and problems.

The Cornell Critical Thinking Tests. These are actually two different tests, Level X and Level Z (Ennis & Millman, 1971a, b, c). They are not parallel forms of the same test, although they are both general measures of critical thinking ability. Level X is meant to be used for testing students roughly from junior high school through first year college age. Level Z is intended for testing examinees of college age on up, although high ability secondary students could also take the test.

The rationale for the test is based upon a concept of *critical thinking* set forth in Ennis (1962). According to the test manual,

    A critical thinker is characterized by proficiency in judging whether:
    1. A statement follows from the premises.
    2. Something is an assumption.
    3. An observation statement is reliable.
    4. An alleged authority is reliable.
    5. A simple generalization is warranted.
    6. A hypothesis is warranted.
    7. A theory is warranted.
    8. An argument depends upon an ambiguity.
    9. A statement is overvague or overspecific.
    10. A reason is relevant.   (Ennis & Millman, 1971a)

Not all of these aspects of critical thinking are covered by each test. Level X does not cover aspects 7, 8, and 9, while Level Z does not cover aspect 7.

In the Level X test, examinees read a story about space explorers on a distant planet and are asked questions as the story unfolds. For example, at one point the explorers are said to be watching a group of beings who are standing around a campfire. In one type of question, the examinees must decide whether the first of the following underlined statements is more reliable than the second, the second more reliable than the first, or whether neither is more reliable than the other.
    A. The mechanic, looking through his field glasses, says, *"They are tan-skinned creatures with furry spots."*
    B. The anthropologist, looking through his field glasses, says, *"They don't have furry spots. They are wearing the skins of animals."*
    C. Equally reliable or unreliable.

The Level Z test is somewhat more difficult and the questions and directions are more complex. For example, after an experiment and its results are described, examinees are asked whether certain additional information, if true, would make the results (a) more certain, (b) less certain, or (c) neither. There is also a section on semantic skills and concepts not covered on the Level X test.

Like the Watson-Glaser test, the Level X test has its problems with induction. It tries to avoid some of these problems by asking whether proposed evidence counts for, against, or is neutral with respect to a certain hypothesis; that is, it only asks in which direction the evidence points. But as in the case of the Watson-Glaser test, one can, by making reasonable assumptions, dispute the keyed answers. We feel that this is the major problem that people writing tests of inductive-reasoning ability must overcome.

Some of the subsections of the Level Z test seem a little short. For example, the section on judging the reliability of observations and authorities is only 4 items long, as compared with 21 items on the Level X test.

The *Watson-Glaser Critical Thinking Appraisal,* the *Cornell Critical Thinking Test, Level X and Level Z* are, in our opinion, the only general tests of critical thinking currently available. But there are two other available tests that claim to measure or appear to attempt to measure general critical thinking ability, and there is a set of "indexes" that might jointly be construed as a critical thinking test.

One test is ***The Uncritical Inference Test*** by William V. Haney (1975). It consists of three stories, each followed by a series of statements (76 items altogether) which the examinee must decide are either "definitely true," "definitely false," or "?" (questionable) based on the information given in the story. Due to the large number of answers keyed "?", a hardened skeptic or "pathological doubter" would receive a high score on this test. (This was a problem with an earlier version of the Watson-Glaser test, also. See Ennis, 1958). The test appears to be essentially a test of whether a person is willing to infer at all beyond what is very explicitly stated, and might better be called a critical uninference test.

The other is entitled ***Logical Reasoning,*** developed by J. P. Guilford and

A. F. Hertzka (1955), designed to assess what Guilford calls the factor of evaluation of semantic implications. Although he claims that this factor is known as critical thinking ability, the test itself consists of forty items, each of which presents two premises of a syllogism and asks the examinee to pick the correct conclusion from four possible alternatives. This seems to be a test of only one aspect of critical thinking, viz., class reasoning.

Then there are several aspect-specific tests of critical thinking ability. The *Cornell Conditional Reasoning Test* (Ennis, Gardiner, Guzzetta, Morrow, Paulus, & Ringel, 1964) and the *Cornell Class Reasoning Test* (Ennis, Gardiner, Morrow, Paulus, & Ringel, 1964) are designed to test the abilities named in their titles. *The Evaluation Aptitude Test,* by D. E. Sell (1952), consists of 36 syllogisms. The test is divided into two parts, one containing neutral items and the other containing emotionally-loaded items. The test is meant to be used, in part, to measure the degree to which emotional bias influences deductive-reasoning ability.

A set of aspect-specific tests has been constructed by the Instructional Objectives Exchange (IOX), formerly associated with the Center for the Study of Evaluation at UCLA. It is an organization that serves as a clearinghouse for instructional objectives and tests designed to measure whether or not students have attained those objectives. Each test (or "index," as they put it) is meant to measure the attainment of one or more stated objectives.

In *Judgement: Deductive Logic and Assumption Finding* (Instructional Objectives Exchange, 1971), seven objectives from the area of informal logic are presented in conjunction with five indexes designed to measure the attainment of those objectives. The *Conditional Reasoning Index* and the *Class Reasoning Index* are meant to measure two objectives each. The work of Ennis and Paulus (1965) is cited as a basis for the items included, but one should note the sense in which the IOX authors use the term 'valid.' The directions ask examinees whether a proposed *conclusion* is valid or invalid, given one or more premises, rather than whether an argument, or line of reasoning, is valid.

Two other objectives and their corresponding indexes deal with assumption-finding. However, the term 'assumption' has several senses and this could cause some problems if one is not alert. Objective 5 in IOX (1971, p. 4) states the following:

> Given a series of statements, each of which is followed by several proposed assumptions, the students will determine whether, within each question set, each of the assumptions listed is necessary to the particular statement.

Assumption Recognition Index I is the test associated with this objective. Objective 6, on the other hand, involves judging whether statements are necessary to arguments, a skill sometimes called "gap-filling". *Assumption Recognition Index II* is the test associated with Objective 6.

After being presented with a statement in the first test and an argument in the second test, the examinee is asked to decide whether a person who offers an argument or statement must accept each of a series of additional statements in order to be "reasonable and consistent." In *Assumption Recognition Index I*, the keyed answers appear to be either (Strawsonian) presuppositions or (Gricean) implicatures, and in *Assumption Recognition Index II*, the keyed answers are either gap-fillers or implicatures. While one might argue that pre-suppositions are in some sense necessary for certain statements and that certain gap-fillers are needed (though not logically necessary) for some arguments (given a context), it is not clear that implicatures are necessary for either statements or arguments. In addition, using the label "assumption recognition" for tests which involve identifying presuppositions, implicatures, *and* gap-fillers may be misleading to the test user searching for a test to cover a homogeneous notion of assumption recognition.

The final objective and its accompanying test, **Recognizing Reliable Observations,** is also based on the work of Ennis (1962). The items in this test seem clear and well-written, but given the number of criteria listed by Ennis for judging the reliability of observation statements, there is some question as to whether the test is long enough (10 items). This test and the other IOX tests could be useful aids for the informal logic teacher, but, as is the case with any test that one has not constructed oneself, they should be examined carefully before they are used. There is no discussion of cut-off scores, nor any argument offered for the representativeness of the items.

The tests mentioned above are the ones we have found that are still in print.* They are contained in a critical thinking test file we are developing at the Illinois Rational Thinking Project, for which we hope to obtain copies of all existing general and aspect-specific critical thinking tests. One project member, Bruce Stewart, has written a collection of reviews of tests in the areas of critical thinking and informal logic (Stewart, 1979), containing reviews of about thirty tests.

Another source for information about tests of all sorts is Buros' **Mental Measurements Yearbook** (1972). Buros lists tests in many areas and also includes reviews of some. The 1972 edition is now somewhat out of date, but a new edition should appear soon.

CONCEPTS OF TEST THEORY *(Appeared Late 1979: Eds.)*

Although the selection of tests of interest to the teacher of informal logic is unfortunately somewhat limited, it is useful to know what criteria are generally used to choose a text. An understanding of these criteria and an acquaintance with their theoretical background can help one get a better overall picture of the process of evaluation of teaching and research in informal logic.

Norm-referenced and criterion-referenced testing. A test that is

---

*Perhaps Sell's *Evaluation Aptitude Test* is out of print now. It is not listed in the most recent Psychometric Affiliates catalogue.

intended to measure the absolute standing of individuals with respect to some standard of performance (often mastery) is generally called by test theorists a *criterion-referenced* or *content-referenced test*. ('Criterion-referenced' is a term introduced by Glaser (Glaser & Klaus, 1962).) Its inclusion of the word 'criterion' is somewhat unfortunate, since 'criterion' has another use in test theory that could cause some confusion. (See the discussion of predictive validity below.)

Someone might alternatively be interested, not in assessing *degree of achievement*, but rather in determining *differences* among groups of students and individual students. A researcher making comparisons would use a test that assesses the *relative standing* of individuals with respect to the possession of a trait or traits. This type of test is generally called by these theorists a *norm-referenced test*. A person's score on such a test is an indication of how well he or she performed as compared with other individuals who took the test. Someone who scores at the 90th percentile of a norm-referenced test has done as well or better than 90% of the people with whom he or she is being compared. Such scores, however, do not indicate whether the level of mastery was low or high. So, for example, someone who scores at the 90th percentile on a test of reading ability may not be a very good reader. He or she is just better than 90% of the group with whom he or she is being compared (the *norm group*).

Since it appears that the same test can be used both as a criterion-referenced test and a norm-referenced test, we propose to change the labels slightly and talk of criterion-referenced test*ing* and norm-referenced test*ing*. This labeling explicitly recognizes the dependence of the distinction upon purpose and interpretation in the given situation. This relabeling relieves us of the burden of classifying every test as one or the other type, a task we have found in practice to be impossible.

At the present time, the bulk of published tests are developed and defended for the purpose of norm-referenced testing. Examples of such tests include IQ tests and college entrance exams, such as the *Scholastic Aptitude Test* (SAT) and the *Graduate Records Exam* (GRE). The *Watson-Glaser* and *Cornell Critical Thinking Tests* might be usable for either purpose, depending on whether they are judged by a test user to be adequate in a particular situation. The IOX tests are designed to be used for criterion-referenced testing. Competency testing, which is becoming popular in elementary and secondary schools, is a type of criterion-referenced testing. (For a discussion of problems with competency testing, in addition to the general problems with criterion-referenced testing that we will mention, see Smith, 1975.)

The fact that a test was designed for norm-referenced testing does not preclude its use for criterion-referenced testing (or vice-versa). However, some knowledge of the theories or models behind a test is helpful in deciding whether a particular test is appropriate for a use one has in mind.

Most of the testing terminology that we shall introduce was originally used in the context of classical mental test theory, which was developed to cover norm-referenced testing. It should be noted, however, that norms are not necessary for the employment of classical test theory (see Lord & Novick, 1968, p. 34, for the assumptions of classical test theory). The key concept in classical test theory *variance,* not norms. In fact, many of the terms of classical test theory can be used when discussing criterion-referenced testing, but it is not clear that all of the terms make sense when so employed.

True scores. Although there are now several types of mental-test score models, classical test theory is what is known as a *true-score model*. On this model, an individual's observed score on a test, X, consists of a true score, T, plus an error score, E. A partly philosophical problem, the nature of true scores, is discussed by test specialists, Frederick Lord and Melvin Novick (1968, pp. 39-44). Lord and Novick also discuss the basic assumptions of classical and other test theories.

Reliability. The concepts of *reliability* and *validity* are very prominent in the literature on testing. By definition a test is *reliable* to the extent that it produces consistent results from one application to the next; and, roughly speaking, a test is *valid* to the extent that it measures (or correctly appraises) what it is supposed to measure (or correctly appraise). These are very rough definitions, but they do give one an intuitive handle on the concepts. Both concepts are problematic in application.

In contrast to discussion of all but two types of validity, discussions of test reliability generally have at least the appearance of precision, as a result of the complex statistical techniques that are employed to investigate the reliability of tests. Despite their complexity, however, these techniques are much easier to deal with than the controversial procedures of test validation. These facts may explain why there is an inordinate amount of emphasis placed on establishing test reliability as opposed to showing the validity of a test. In choosing tests, one is likely to encounter data on test reliability rather frequently.

As defined above, a test is reliable to the extent that it produces consistent results from one application to the next. This is similar to what one would expect in a theory of measurement in the physical sciences. A ruler is reliable to the extent that it produces a consistent set of data from one application to the next. To determine whether a ruler was reliable, one might repeatedly measure the same thing. A reliable instrument would produce very close readings on the repeated measurements. For some types of tests covered by educational test theory, such as physical or motor skills tests, this notion of reliability will suffice. But for most educational tests, reliability cannot be determined in terms of direct comparison of repeated measurements of the same individual using one instrument. This is partly because human beings are often changed by the measurement process itself. The act of taking a test can affect the trait being measured by the test. Consequently, one would not expect consistent scores on repeated measures. In fact, one might find *artificially consistent* scores, since examinees sometimes remember their original responses when taking the retest.

In an attempt to surmount this problem, test theorists employ the notion of a parallel test form. *Parallel test forms* are defined as tests that produce parallel measurements. Two measurements are said to be *parallel measurements* if each individual's true score on the two measurements is the same and if the variance of the error scores for the two measures are equal (that is, the error scores for the two measures are spread out to the same extent). What this notion does for test theorists can be seen in the following quote from a standard text in this area, ***Statistical Theories of Mental Test Scores*** by Lord and Novick (1968, p. 48): "Thus parallel measurements measure exactly the same thing in the same scale, and, in a sense, measure it equally well for all persons."

The reason that parallel measures are used in discussions of reliability is that one common method of defining reliability involves the unobservable quantity, an individual's true score; that is, reliability is defined as the squared correlation between observed score (X) and true score (T): ρ^2_{XT}. It deductively follows from the assumptions of the theory, however, that $\rho^2_{XT} = \rho_{XX'}$ where X and X' are parallel measures and are potentially observable. So the concept of parallel forms is helpful in developing a usable theory of reliability.

It is very often judged too difficult, too time-consuming, or too expensive to develop a parallel form for a test. When this is the case, there is a third, widely-used approach to estimating the reliability of a test: estimating the internal consistency of the test. These estimates use only a single test form to estimate reliability. One such procedure is the *split-half method*. In this method one first divides the test into two halves that are assumed to be parallel. (There are, of course, many ways to split a test. How the splitting is done depends upon the nature of the test.) The scores on the two split-halves are then correlated. This does *not* produce an estimate of the reliability of the original test, but of a test only half as long. To estimate the reliability of the original test, one uses a formula, the "Spearman-Brown" formula, that estimates the reliability of a test that is longer than a given test.

In addition to split-half internal consistency, there is another common single-form approach to estimating the reliability of a test. This approach we shall call the *multiple internal-consistency approach,* since it relies on the extent to which *all* of the items intercorrelate with each other. The Kuder-Richardson formulas (KR-20 and KR-21) are commonly used ways of estimating multiple internal consistency.

There are other methods of estimating reliability by looking at the internal consistency of a test, and one can find a discussion of these approaches, as well as a cogent, but somewhat technical, discussion of the concept of *reliability* by Julian C. Stanley in a book chapter entitled, "Reliability" (1971).

There is a significant problem in the use of multiple internal-consistency reliability estimates. *Informal logic* and *critical thinking* are probably heterogeneous notions, but multiple internal-consistency reliability gives higher ratings to homogeneous tests. Hence in building tests to produce high multiple internal-consistency reliability there is the tendency to eliminate items that do not correlate highly with the rest—even though such items may be very good indicators of some feature that is not highly correlated with the other features of these heterogeneous notions. Such items tend to be eliminated by test makers interested in making the reported numbers look good, and the reason that such items are eliminated is simply that they are non-conformist.

This problem of using multiple internal-consistency as a substitute for the original notion of reliability (consistency of repeated applications) is not peculiar to the domain of testing in informal logic and critical thinking. It permeates most mental testing by highly respected organizations. Informal logicians are especially suited to guard against the resulting invitation to equivocation in defense of a test. We urge their sharing of this insight with others less sensitive to equivocal arguments.

Validity. It is not enough for a test to give consistent scores; it also must measure what we desire to measure. The process of determining whether a test measures what it is designed to measure is called *test validation*. When sufficient evidence has been accumulated to support the claim that the test measures a certain variable, the test is said to be a *valid* measure of that variable. Actually, this way of speaking is slightly misleading, although it is often encountered. Cronbach (1971, p. 447) urged that it is not the test instrument *per se* that is valid, but the interpretation of data arising from a particular use of the test. A single test can be used in many ways (e.g., research, placement, job-screening, grading, etc.). Some interpretations for a particular use may be valid; other interpretations of data from the same test used for different purposes may not be valid.

Although Cronbach's point is an important one, most discussions of tests are still carried on with references to the "validity of a test." One should understand such locutions as essentially incomplete expressions. A test's validity must be understood as its ability to measure or be a sign of one or more specified things when it is given under particular conditions. These additional qualifications must be kept in mind if one encounters talk of the validity of a test instead of talk of the validity of interpretation of test scores.

Despite the appearance of precision given by the statistical superstructure of test theory, many of the central concepts in this field are not at all precise or well-clarified. *Validity* is one such concept. There are actually several somewhat loosely-related concepts which come under the heading "validity." We shall briefly characterize five of the concepts that one is likely to encounter in discussions of testing.

A test is said to have *face validity* if it appears to be a valid test. According to the American Psychological Association's **Standards for Educational and Psychological Tests** (1974), face validity is the "mere appearance of validity" (p. 26). Most test experts feel that face validity is illegitimate, but we do not see how to get along without it. It appears to be an essential element of content validity.

To explain their notion of content validity, test theorists introduce the concept, *universe of behaviors*. Such a universe consists of a set (possibly infinite) of behaviors that a student should be able to exhibit if he or she has grasped certain concepts or mastered certain skills. This concept, *universe of behaviors,* is ripe for philosophical inquiry. Although we shall use it in presenting established theory because it is part of established theory, we have many reservations about its use.

We often design tests to determine whether our students have learned what we intended to teach them. For example, after a unit on propositional logic, we want to be able to determine whether our students learned something about that topic. We cannot, of course, test for all possible "behaviors" we would expect a successful student to be able to exhibit. We are expected to try to get a representative sample of the universe of behaviors for which we want to test. How one can (without leaning heavily on face validity judgments) identify a representative sample of a critical-thinking universe of behaviors is, unfortunately, unclear. However, a test is said to have *content validity* to the extent that we actually did choose a representative sample. It does seem that a test on propositional logic that only asked questions of the following form would not have much content validity:

Is the following argument valid?

If r, then s.
r.
_____
Therefore s.

Assuming that the only difference among the items is the use of different single letters in the place of 'r' and 's,' the items do not appear to call for a representative sample of propositional logic behaviors, in whatever way we choose to interpret the word "behaviors." But that judgment appears to be a face validity judgment. If not, then we would have had to have a way of describing and identifying the things in a (infinite) domain of propositional logic behaviors and of drawing a random or systematic sample from that domain. This describing, identifying, and sampling makes no sense to us, but we invite other philosophers to work on the problem.

The resolution of this difficulty is important since the clarification of the concepts of *face validity* and *content validity* is essential for the further development of theories applicable to criterion-referenced testing. Such testing is of paramount interest to those who wish to evaluate the extent to which students have mastered the content of informal logic courses.

Two further types of validity, distinct from the previous two but related to each other, are *predictive validity* and *concurrent validity.* In many cases, a tester is interested in estimating the value of a certain variable from the score on a test. For example, college admissions officers would like to estimate a candidate's freshman grade point average, given the candidate's score on an entrance exam, such as the SAT or ACT. The variable to be estimated is called the *criterion.* (This kind of criterion should not be confused with that in criterion-referenced testing. Predictive and concurrent validity are not usually concerns in criterion-referenced testing.) A test has *predictive validity* if knowing a subject's score on the test enables us accurately to predict the value of the criterion. The difference between predictive and concurrent validity lies in the temporal relation between the test score and the criterion. One speaks of *concurrent validity* when one is interested in a subject's standing on the criterion at the time of test administration, while predictive validity is the concern when one is interested in the subject's standing on the criterion at some future time.

Concern with predictive validity at one time dominated test theory, although at the present time more attention is being given to construct validity than was given to it in the past. One is investigating the *construct validity* of a test when one attempts to confirm or disconfirm that a test measures some hypothetical, unobservable psychological construct. *Intelligence, anxiety,* and *critical thinking ability* are examples of such constructs. The claim that a test measures intelligence, we feel, is a claim about the construct validity of a test, as is the claim that a test measures critical thinking ability. (Behaviorists, who reject the idea of construct validity, would of course not agree.)

The investigation of construct validity can be viewed as the process of placing the specified construct in the context of some larger theory, and ascertaining the acceptability of the theory, the role the construct plays in the theory, and the relationship between the test and the construct.

> Construct validation is the process of marshalling evidence in the form of theoretically relevant empirical relations to support the inference that an observed response consistency has a particular meaning. (Messick, 1975, p. 55)

Construct validity is viewed by some as a broad concept that encompasses or subsumes all other types of validity.

The model employed in construct validation is the neo-positivist theory of confirmation as set forth by Carl Hempel (1965, 1966). Statements about the construct in question are located in a hypothetico-deductive system. Predictions are deduced from the system and either confirmed or disconfirmed. The construct validity of the approach to interpretation of test scores is supported to the extent that the predictions are confirmed and to the extent that the predictions depend upon the relationships among the test, the construct, and the other elements of the system (which are also constructs). Philosophers are well-acquainted with the extensive and powerful criticisms of this model, but such criticisms have as yet had little effect on the use of hypothetico-deductive models in test theory. (This is not to say that test theorists are completely unaware of the problems involved, cf. Cronbach, 1971.) A standard philosophical problem connected with construct validity is the nature of the constructs being investigated. For example, what (if anything) is being referred to by the term "critical thinking ability"?

Judging tests. How should one use the concepts discussed above when judging an available test? That depends to a great extent on the purpose for which one is using the test. Some remarks about general cases, however, can provide some guidance. One must take care when using the information provided about tests, since, as we have to some extent indicated, there are problems involved with theory behind tests and their interpretation.

One very general problem involved in making judgments for criterion-referenced testing is the source of the vocabulary which is used to talk about tests. Classical test theory was developed to cover norm-referenced testing. Some of the terms which are appropriate to use when discussing norm-referenced testing are not clearly applicable to criterion-referenced testing. There is as yet no theory of and vocabulary for criterion-referenced testing comparable to the theory and vocabulary which have been developed for norm-referenced testing, although progress has been made in this area during the last decade. (See Hambleton, Swaminathan, Algina & Coulson, 1978.) Nevertheless, some test theory concepts seem applicable to both types of testing.

Reliability as a basis for judgment. High reliability seems to be desirable for any test although, as indicated earlier, the commonly-used internal-consistency formulas for estimating reliability are misleading for nonhomogeneous tests. There are no firm requirements for reliability coefficients, although some have been suggested. A widely-quoted set of minimums was set forth by Kelley (1927):

(a)	To evaluate level of group accomplishment	.50
(b)	To evaluate differences in level of group accomplishments in two or more performances.	.90
(c)	To evaluate level of individual accomplishment.	.94
(d)	To evaluate differences in level of individual accomplishment in two or more performances.	.98

These figures of course depend on the assumptions Kelley made about required fineness of discrimination and the acceptable chances of going wrong. They also partly explain why many professional testing people are so devoted to obtaining high reliabilities: Here we have some "objective" standards, and there are test development procedures that generally will enable one to meet these standards—at a cost. The cost might be (1) excessive testing time, (2) excessive demands on the time of experts, (3) triviality of items, and (4) neglect of important features of the trait(s) for which one is testing. (Remember the pressure for homogeneity of items resulting from the use of internal consistency formulas for reliability estimation.) To the extent that we accept these last two costs, a frequent occurrence, we get a reliable, but invalid test.

Heavy reliance on reliability is also related to the fact that early test developers were often interested in predicting the standing of a subject on some criterion. They were concerned with predictive validity. According to Lord and Novick, reliability can be viewed as predictive validity with respect to a parallel test (1968, p. 63). Consequently, reliability was seen as part of the only concept of validity thought to be important, criterion-related validity.

One way to increase multiple internal-consistency reliability is to secure item homogeneity. Another, according to classical test theory, is *simply* to increase the number of items on the test. To illustrate this phenomenon, consider the *Cornell Critical Thinking Test, Level X* and the *Cornell Critical Thinking Test, Level Z*. Depending on the group from which the data was collected, the estimated reliabilities of the tests range from .77 to .87 for Level X and from .55 to .77 for Level Z. However, Level Z is only 52 items long, as compared with 71 items on Level X. Using the Spearman-Brown formula, one can show that if Level Z were as long as Level X, its reliability estimates would range from .62 to .82, which is closer to the range for Level X.

Most currently-used procedures for estimating reliability, even including split-half methods, do not capture the full-blooded notion of reliability. They are based on one test administration. Consequently instability of measurement over repeated administrations is not taken into account. (For a more detailed explanation of this problem, see Cureton, 1965. Also, see section F of the American Psychological Association's *Standards for Educational and Psychological Tests* (1974).)

The concept of *reliability* must be cautiously applied to criterion-referenced testing. Some techniques which can be used to increase the reliability for norm-referenced testing are not appropriate for criterion-referenced testing. For example, when revising a test, the reliability can be increased if one retains items with a "difficulty index" of about .5, meaning that the proportion of examinees obtaining the correct answer is .5. (This index is misleadingly named. It might better be called the "ease index," as suggested by Ahmann & Glock, 1958.) If instruction has been effective, the difficulty (read "ease") index should be high for items on a test for criterion-referenced testing, meaning that a high proportion of students should answer the items correctly. A test maker who aims for items with a .5 difficulty index will then be forced to construct overly difficult, recondite, or nit-picking questions.

In summary, there are a number of traps facing someone pursuing high reliabilities, and in accepting the judgments of others who pursue high reliabilities.

Validity as a basis for judgment. In almost all cases of interest to the teacher or researcher in informal logic, one must also ask whether a test is valid. As a first step in judging the validity of a test, one should examine the description of the test that should be in the manual to see whether the test comes close to what one seeks. Then, if it appears worthwhile to go on, one should scrutinize the items very carefully. The best way to do this is to take the test under the prescribed conditions and check one's answers against the key, seeking for explanation and resolution of any discrepancies. After going through this process, you will have a fairly good idea about the extent to which the test does what you want it to do. A judgment based on such an inspection would be a judgment about the so-called "face validity" of a test. Going through these steps makes good sense, even though face validity is a disreputable notion in the eyes of many test theorists, making this low regard somewhat puzzling.

Judging a test for its content validity, as defined above, requires that one adopt and employ the concept *universe of behaviors*. As indicated earlier we shall provisionally do so for the purposes of applying this approach.

Judgments about the content validity of a test should be aided by the examination of the test rationale that should appear in the test manual. This rationale should somehow help one identify all the members of the universe of behaviors, so that one can then decide whether the test items call for a representative sample from it. How one actually does all this we do not know. In one sort of actual practice it appears that content validity is established by making the topics in the rationale quite specific and, if possible, by transforming them into types of behavior to be exhibited in types of situations (rather than into specific items of behavior*). This list of types of behaviors is called a table of specifications. Then face validity judgments are made (though they are not called face validity judgments) about the item-produced behaviors' representativeness of the types of behavior in the table of specifications.

A second procedure for establishing content validity in actual practice is to gather a large number of items that an expert judges (another face validity judgment) to call for behaviors that are representative of types of behavior desired. Then a random sample of some sort (or a systematic sample) is drawn from the item pool, and the test consists of this sample.

Note that both of these procedures for establishing content validity do in fact lean heavily on face validity judgments. Content validity, in the only ways we can conceive of its pursuit, consists of organized systematic ways of utilizing the face validity judgments of experts. Both of the content-validity procedures we have outlined can be followed by someone building a test of logical competence, and can be evaluated for their care and quality by consumers of such tests.

These processes of judging pure face validity and content validity are applicable to any critical thinking test, whether for norm-referenced testing or criterion-referenced testing, and whether the test was originally constructed for norm-referenced or criterion-referenced purposes.

A problem that has still not received much attention in the literature is the problem of making judgments about desirable levels of performance for criterion-referenced testing. What level of performance should be considered evidence of mastery? This is a difficult question to which developing theory does not yet have an answer, even though the test user often seeks an answer to this question. (See Popham, 1971, for a sympathetic discussion of problems in the theory of criterion-referenced testing.)

Establishing the construct validity of tests is a difficult task, in part because *construct validity* in itself is not a crystal-clear notion. Much more attention has been given to construct validity in the past several years than was given to it in the early days of the development of test theory. But

---

*This distinction is between behaviors' being dispositions and their being performances.

problems still remain.

As described above, the process of making a case for the construct validity of a test consists of showing how the construct fits into some larger theory. One way to do this is to show how scores on the test in question are related to other variables. So, for example, one would expect that critical thinking ability, since it involves judgments about statements, would be moderately related to reading ability. Many test manuals offer lists of correlations of the test in question with other variables. But such a list by itself does not establish the construct validity of a test. One must show how the correlations would be expected to follow from a theory in which the construct in question in embedded.

One important place to look for evidence of construct validity is in the relation between a test and other closely related measures. For example, one would expect a high correlation between tests which claim to measure critical thinking ability. Such "convergence of measures" gives some support to the construct validity of all of the tests involved. Lack of agreement could mean several things: a poorly constructed test, differing conceptions of critical thinking, unshared prerequisite familiarity with the subject matter, etc.

One might also expect the constructs measured by a particular test to be *unrelated* to certain other constructs, that is, one should be able to discriminate between unrelated constructs. Tests measuring unrelated constructs should be weakly correlated or uncorrelated (see Campbell & Fiske, 1959).

Construct validity arguments for existing critical thinking tests are either weak (The Cornell Tests and the Watson-Glaser test) or nonexistent (the others). This is a comment about the *arguments* for construct validity, not about the construct validity of the tests.

There are certain positions of which one should be aware when reading discussions about construct validity. One may encounter those who demand a reductionist operational definition of each construct. The strict operationalist does not view such a definition as providing *a* method of measuring the construct in question, but thinks that each test defines a different construct. There are many criticisms of this view (for example, Ennis, 1964), and even neo-positivists such as Hempel (1961) regard such a position as too rigid, but one frequently encounters this position in discussions of educational testing. Holders of this position are opponents of the use of construct validity in test appraisal (for an example, see Bechtoldt, 1959).

Another position that one occasionally encounters holds that high correlation implies conceptual identity. That is, if two tests correlate highly then they measure the same thing. Cronbach proposes a counterexample to this position (1969): Comprehension of physical laws will correlate highly with scientific reasoning ability, but this does not mean they are identical. It may simply be the case that, at present, the best curricula do a good job on both and the worst do a poor job on both.

Construct validity questions are usually associated with norm-referenced testing. Some experts in testing feel that one need not consider construct validity when assessing tests for criterion-referenced testing. However, there has recently been criticism of this position. The literature in this area is just now developing and little can actually be reported at present, but it is an area that merits watching and participation by philosophers as it develops.

Sometimes predictive and concurrent validity will be of use to informal logicians, and when they are, the goal is high correlations between the test and the criterion. Generally correlations between scholastic aptitude tests and levels of later subject matter achievement run about .5 (this is predictive validity); correlations of tests with other tests that are testing for the same thing go up to .8 when the tests are fairly similar (this is concurrent validity when the other tests are administered at roughly the same time). These numbers might serve as rough guides for what one can expect. Statistical significance of correlations is generally of little interest for predictive and concurrent validity, since that standard is too easy to satisfy with a sample of any reasonable size.

The major problem in using predictive and concurrent validity in evaluating informal logic tests is that of finding a criterion that can justifiably be assumed to be better than the test in question.

The suggestions given above for assessing tests are not meant to be exhaustive. Many other considerations enter into the choice of a test (e.g., time limits, cost, reading level).

CONSTRUCTING TESTS

After examining available informal logic tests, one might conclude that none are appropriate for the purposes at hand. At this point a natural move would be to consider the task of constructing one's own test.

Constructing a good multiple-choice test is no easy task. It cannot be accomplished at one sitting, since, ideally, test construction involves several distinct time-consuming steps. One might object that the time and effort involved would not be justified if one just wants to make up a mid-term exam. While we might not ordinarily undertake a grand project in such a case, nevertheless, attention to the procedures outlined below can improve the quality of many tests.

As one would expect, the procedures for constructing instruments for norm-referenced and criterion-referenced testing are somewhat different, although one instrument might serve both purposes. Each of the following list of procedures is an edited and abridged version of a presentation in Sax, 1974. If you follow these procedures, it is essential to ask frequently whether what you are doing makes sense. Mechanical rule following is dangerous, but an easy trap to fall into. If test specialists are employed, the informal logician must monitor the process closely.

Constructing tests for norm-referenced use. The following steps describe the procedures one would usually follow in constructing a test for norm-referenced use.

1. Test rationale and objectives are determined. This serves as a foundation for the writing of items. It also serves as part of the case for the face, content, and construct validity of the test. The content of the test is determined by the nature of the field in some tests and by the type of objective to be tested for in others. Subject matter experts should be involved.

2. Next, items are written for the test. More items are written than will be included in the final form of the test. Sometimes several preliminary versions are constructed. Although the multiple-choice format is most often used, other formats are possible (see "Test format" below).

3. The items are administered and the results are analyzed. It is desirable to give the proposed items to a fairly large and varied sample of the population for which the test is designed. For some widely used tests, such as the SAT, tens of thousands of examinees take the trial tests, but much smaller samples can be used. Most colleges and universities and some high schools now have computer facilities which give detailed item analyses for machine-scored tests.

Two standard results of an item analysis are the difficulty index (described earlier), and the discrimination index. The discrimination index is an attempt to indicate how well an item distinguishes between two groups of people otherwise identified. Often these groups are the top and bottom groups on the total score on the test; if so, then there is a danger of overemphasizing homogeneity and neglecting important aspects of informal logic, if any, that do not correlate highly with the ones that dominate the test. Seeking high discrimination indices based upon total score helps achieve high internal-consistency reliability estimates, a trap mentioned earlier. But in any case items with low discrimination indices should be carefully scrutinized. Often the cause is a problem in the wording of the item. Poorly worded distractors (supposedly incorrect alternative answers) that should not be scored "wrong" can often be detected by item analyses that indicate how groups selecting each distractor performed on the total test.

For norm-referenced testing, difficulty indices of .5 are often sought because that is a good way to spread people out on a continuum. Dangers of using this standard for criterion-referenced testing were mentioned earlier, and they apply to some extent to norm-referenced testing as well. We might succeed (by following this procedure) in spreading people out on a continuum, but the continuum might, as a consequence, be of little interest.

In any case, the results of item analyses, one should remember, apply at best to groups similar to the group that took the test. They might not apply at all to a different group. A danger here is to do an item analysis using a group that has received no instruction in informal logic, and then to use the test on a group that has had considerable instruction in informal logic. Opportunities for distortion abound. Reliability estimates should be computed, and validity evidence should be considered.

4. The final test form is constructed. Factors such as time limits for test taking influence the number of items included in a form.

5. The final form is administered to another large and varied group of examinees and normative data are generated for use in subsequent administrations. Widely used tests are continually being revised and norms are frequently updated.

At this point, one would have a test that is norm-referenced, but not necessarily a *good* test. If items are chosen with reliability in mind after step 3, the reliability of the test is likely to be high. Even if evidence regarding face, content, and concurrent validity is present, predictive and construct validity would still need to be determined. It should be apparent that the construction of a good test for norm-referenced purposes could take several years.

Constructing tests for criterion-referenced testing. In constructing such tests, one follows procedures similar to but not identical with norm-referenced procedures. One should also keep in mind that these tests do not have the backing theory which norm-referenced tests have.

1. A general test rationale is prepared.

2. The universe of behaviors to be covered by the test is specified. These specifications indicate what a student who has mastered the universe should be able to do, although it is not clear whether the "behaviors" are to be dispositions or performances.

3. Test items are written which conform to the specifications of step 1. How to do this is not clear, because in most content areas, the nature of the relationship be-tween the items and the universe of behaviors is not clear. Be that as it may, one next makes (ideally) a random selection from all such possible test items, but this is usually not possible since most universes subsume an infinite number of items. Instead, one might try to assure that the sample of items selected is representative by comparing the items with the universe specifications (a face validity judgment).

4. If one has to choose from among the items selected in step 2 (to adjust the test for proper time length, for example), those items that discriminate most clearly between those who have had instruction and those who have not are usually preferred, other things being equal.

5. Standards of competence are determined. There is controversy over whether this step can or should be taken. Although many criterion-referenced tests have cut-off scores, Glass (1978) argues that procedures used in determining cut-off points are indefensible.

6. The test is administered under conditions that conform to the universe specifications (i.e., if the universe of behaviors deals with written criticism of *written* arguments, an oral test would not be appropriate).

7. Student performance is assessed by comparing test results with the specified standards of competence. One checks to see whether the ratings of the students make sense.

As with norm-referenced testing, these procedures lend themselves to continuing test revision and improvement over time. Items that discriminate most clearly between those who have and have not mastered the material are retained and other items are scrutinized for deficiencies. Various forms of a test can be developed by taking different samples from the domain of test items conforming to the universe specifications.

The controversy over step 5 points to a popular misconception about the nature of criterion-referenced testing. Some people believe that a test for this purpose is essentially one which classifies an examinee as competent or incompetent with respect to some skill. This is not widely viewed by test theorists as a necessary characteristic of such a test, although some theorists view it as highly desirable for practical applications. What is necessary is that the score be directly interpretable in terms of behaviors or performances. Deciding what *level* of performance constitutes competence is an extra step.

At least some of the procedures outlined here can be helpful even where a teacher simply wants a mid-term exam that will only be used once. At least stating the rationale and specifying crucial behaviors (either dispositions or performances) can be helpful in thinking through the test specifications.

Test format. An informal logic teacher interested in constructing an achievement test is faced with the problem of deciding what type of item or items to construct, e.g., multiple-choice or essay. There is an amazing variety of item forms used in tests, but for discussion purposes, we will classify them into three main types: multiple-choice, short-answer, and essay. These types are distinguished by the latitude a student has in constructing an answer. Multiple-choice tests allow a student to pick only from specified possible answers. There are many kinds of multiple-choice forms available (true-false, matching, multiple-response, etc.) and many ingenious variations have been invented to measure a wide range of objectives (see Anderson, 1972, and Wesman, 1971). The short-answer item, such as a sentence-completion item or identification question, allows students more freedom in that they must supply the answer themselves. Students are limited to some extent by the space allowed for the answer and the necessarily limited nature of the question asked. The essay or open-ended answer allows students a great deal of freedom in choosing what they believe to be a good answer.

The problem of the type of item to employ is a thorny one. Test theorists have traditionally favored multiple-choice items, the type for which the concept, *universe of behaviors,* is best adapted. Such items are easily and inexpensively scored and are not susceptible to errors of measurement caused by inter-grader disagreement. (But comparable errors slip in through the writings of the item and the directions—which always leave room for differing interpretations by examinees.) The data produced by such tests can be analyzed by the sophisticated statistical techniques available to test theorists. On the other hand, students might recognize a multiple-choice alternative as correct when they would not have been able to recall the answer had they been asked a short-answer question.

Essay questions require even more effort and ability on the examinee's part since the structure and content of the answer must be supplied by the examinee. In many cases, informal logicians will want to assess the type of knowledge that essay questions seem best suited to assess. Unfortunately, the concept, *universe of behaviors,* is especially problematic with essay questions.

Since the type of item one chooses to construct depends on the type of knowledge one is trying to assess, it could be useful to have some classification scheme for types of objectives and "behaviors." Benjamin Bloom and his associates have developed a popular scheme for classifying cognitive educational objectives (Bloom, Engelhart, Furst, Hill, & Krathwohl, 1956). The hierarchical list of terms developed by Bloom, *et al.,* is as follows: knowledge, comprehension, application, analysis, synthesis, and evaluation. Although widely used in the literature on education, this list embodies a host of philosophical and other problems. They are in the list itself and its application to testing. The simple problem of classifying an objective (say, "ability to identify an unstated assumption") gives one doubts about this list. Bloom, *et al.,* actually classify this objective under 4.10, "Analysis of Elements," but it is not clear why they place it there instead of somewhere else.

Although the traditional test-theory view has been that multiple-choice items are to be preferred whenever possible, some writers have recently proposed that, for epistemological reasons, multiple-choice items are the least desirable type. Hugh Petrie (in press) has argued that we should view a test as the introduction of a disturbance that the examinee will correct if the desired achievement has been attained. By limiting the possible responses to a test item (the disturbance), we limit possible novel responses that would also counteract the disturbance. New theories proposed by Petrie and others will no doubt have an influence on the future of testing and evaluation.

USING TESTS AND OTHER TECHNIQUES IN EVALUATING INFORMAL LOGIC: STUDENTS, COURSES, AND CURRICULA

Even the most well-constructed test is not worth much if it is not used properly. A test may give us some information about the performance level of our students or the effect on test scores of different approaches to teaching informal logic. But we do not give tests *just* to obtain such information. We use the information to make judgments about student achievement or the merit of innovative teaching methods. When we do this we are engaged in the process of evaluation.

There are many different things that we evaluate in education. They cannot all be evaluated the same way. Even in any one particular area, there is not universal agreement about how evaluations should be carried out. Nevertheless, we can offer some advice on the use of tests and other techniques for the purpose of evaluation.

Testing and Evaluation. Tests are certainly the most widely used instruments in evaluation studies. Although there are many legitimate uses of tests, they can also be misused. A carefully prepared plan for an evaluation study can help guard against the misuse of tests. We shall discuss the use of tests to assess student performance and the employment of tests in various experimental designs. Tests can also be used for other purposes (e.g., placement), but the following topics will probably be of greater interest to those interested in teaching and research in informal logic.

The use of criterion-referenced testing in evaluation. Criterion-referenced testing is the assessing of a student's mastery. We cannot endorse the current attempt to put this purpose in terms of a universe of behaviors, but feel that no matter how we conceptualize the basis, criterion-referenced testing of some sort is useful.

One may use the test results to assign grades, or to determine which students should advance to the next unit of study and which should remain behind for more work. In either case, one must face the problem of specifying what level of performance indicates mastery of the material studied (or what levels of performance correspond to a certain degree of mastery). Experts in this area of test theory have no help to offer us on this problem, but it is essential that the person who does decide be thoroughly familiar with the test rationale, its items, and the subject matter of informal logic.

The use of norm-referenced testing in evaluation. If one is interested in comparing the relative standing of groups (e.g., informal logic class vs. traditional class) on some variable (e.g., critical thinking ability), then one employs norm-referenced testing. In selecting a test one must determine whether the face or content validity and construct validity of a test under consideration for use as a measuring instrument are appropriate for one's own purpose. It is here that a list of written course objectives would be helpful, for the test specifications for a particular informal logic test might not match the course objectives. A test may, however, measure some things considered important in the course specification and could therefore serve as a partial measure of the constructs under investigation. Unfortunately, people all too often rely exclusively on the title of a test for information about what the test measures. There is no substitute for a careful, critical examination of a test and its manual.

Experimental design. Just as important as choosing a measuring instrument is the choice of an experimental design. Even the best instruments cannot produce useable data unless a proper testing schedule is followed. We will briefly consider some of the more popular designs. The reader is encouraged to examine more thorough treatments of this topic, such as Campbell & Stanley (1963), Winer (1962), or Wiersma (1975).

There is one "design" which most experts do not consider a design at all. In this "design," the results of a pretest and a posttest on an experimental group are compared. (A *pretest* is a test administered before a treatment. A *posttest* is a test administered after a treatment. A *treatment* is any deliberately-introduced change in the environment of the group under investigation. For example, instruction in informal logic would be a treatment.) The problem here is that even if the posttest scores are higher than the pretest scores, one has no reason to attribute this inference to the treatment. Any number of other factors (e.g., maturation, familiarity with the subject-matter induced by the pretest, etc.) could be responsible for the higher scores.

In order to draw meaningful conclusions from the preceding experiment, we also need a control group with which to compare the experimental group. By "control group" we mean a group that is supposed to have the same characteristics as the experimental group except that it does not receive the treatment. The preferred method for obtaining equivalent groups is to choose both groups randomly from the population under study. (There is some disagreement among theorists about whether randomly selected groups are equivalent by definition or whether they are only highly likely to be equivalent. The former position seems to be the commonly-accepted view. Thus this concept of *equivalent groups* employed by test theorists and statisticians differs from the ordinary concept.)

The simplest type of experimental design is the posttest-only control group design in which a posttest is administered to the control group and experimental group. One then looks for a significant difference between the mean scores of the two groups. This design is simple to set up and is not as widely used as it could be. On the negative side, the statistical tests involved are not as powerful as those used in some other designs, and there is often a lingering suspicion that the groups were not equivalent at the beginning, despite the random-selection process.

The most popular true experimental design is probably the pretest–posttest control-group design: In this set-up, both the control and experimental groups are administered a pretest and a posttest. One often-used strategy for analyzing the results is to compute gain scores (the difference between pre- and posttest scores) and to test the difference between mean gain scores for each group for statistical significance. This approach is challenged by some experts, however (see Cronbach & Furby, 1970). Analysis of covariance is now one of the recommended procedures for analyses of data from this design. Roughly speaking, analysis of covariance attempts to compare posttest scores while statistically holding pretest scores (and perhaps other variables) constant. This design gives one a check on group equivalence through a comparison of pretest scores. However, it requires twice as many test administrations as the posttest-only control-group design and there is a problem with attempting to generalize to the unpretested population.

One big stumbling block to the use of true experimental designs is the difficulty in arranging for random assignment of individuals to groups. Most institutional settings are not flexible enough to permit randomization, and, in some cases, there are also ethical and political problems with this manipulation of subjects. In such cases, which in educational institutions is most cases, researchers turn to *quasi-experimental designs*, which are similar to true experimental designs, but differ in a few important respects.

The most widely used quasi-experimental design is the "nonequivalent" control-group design. This design resembles the pretest-posttest control-group design except that subjects are not assigned to groups at random. The groups are taken as they are found in some institutional setting. This means that extraneous factors that influence the selection process may turn out to be responsible for any significant differences which are found. This possibility must be carefully considered when weighing the evidence collected. If one has information about relevant characteristics of the subjects, this can sometimes be taken into account in the statistical treatment of the data by means of techniques such as analysis of covariance. Since this design is often the only one available, readers interested in research that will be conducted under conditions that preclude the use of true experimental designs should consult more detailed treatments of this and other quasi-experimental designs (e.g., Campbell & Stanley, 1963, Kerlinger, 1964, or Airasian, 1974).

Regardless of the type of experimental design chosen, one problem that any researcher must face is generalizing from a sample to a population. Unfortunately, the populations from which samples are drawn in educational research efforts are almost never the populations over which it would be desirable to generalize. For example, one might draw random samples (for a control group and an experimental group) from all the students taking informal logic during a particular semester for the purpose of evaluation of a certain method of teaching logic. The population to which one would like to generalize is that of all informal logic students, including next year's group. But the population from which the sample was drawn was much more restricted. (Each member of a population must have an equal chance of being selected in a random assignment.) Sometimes arguments are offered for the typicality of groups chosen in an attempt to generalize over a larger population (Campbell & Stanley, 1963, discuss this difficulty, calling it the problem of "external validity").

Most commercially-available tests contain tables of norms with which one can compare experimental groups or individuals. Such comparisons are useful for suggesting hypotheses about differences between, e.g., national norms and locally-collected data. They can also give an individual an idea of how he or she stands compared to a norm group. The more accurately the norm groups are described, the more readily one can choose the appropriate comparison group. One should not, however, view normative data as a substitute for control-group data collected by the experimenter.

Statistical significance. In educational research, a result that, given the assumptions, is the sort that could have occurred by chance less than five (sometimes one) times out of a hundred is generally deemed to be statistically significant (this is known as the .05 level of significance). Beware of this approach. With very large groups statistical significance can be attributed to differences that are for practical purposes very small. It is therefore good practice to ask about statistically-significant differences whether they are also practically-significant as well. This requires that one immerse oneself in the situation and inquire about the economic and human cost of producing a given difference, and about whether the difference produced is large enough to be concerned about.

Other Approaches to Evaluation. Thus far we have discussed tests and their use as evaluation instruments. The devoting of a large proportion of this paper to tests and test theory reflects the extent to which this approach to evaluation dominates education at the present time. Are there any other approaches to evaluation?

The answer to the preceding question depends on the extent to which the testing model can be extended to cover everything to be evaluated by educators in general and informal logicians in particular. According to some test experts, most of the efforts expended in evaluation projects should be directed toward the construction and perfection of good tests. For them, evaluation means measurement, and measurement means the use of some instrument covered by some test-theory model.

Responsive evaluation. For some evaluators, however, testing is not the whole of evaluation or even the most important part. One group that employs a somewhat different approach to the evaluation of educational programs and materials is the Center for Instructional Resources and Curriculum Evaluation (CIRCE) at the University of Illinois at Urbana-Champaign, directed by Robert E. Stake. Stake is suspicious of traditional evaluation methods since they are what he calls "pre-ordinate" (Stake, 1967, 1976, Stake & Hoke, 1976). That is, they depend on prespecified notions of how a successful program or course must appear and be. By looking only for certain kinds of results (usually in terms of test scores), traditional evaluators may overlook things that would be considered just as valuable as the prespecified objectives, if they were noticed.

CIRCE evaluations tend to include a great deal of narrative or "portrayal" material gathered by observers. These observers make note of things they feel are important and judgments of students, teachers, parents, and school administrators. Rather than evaluating a program or course strictly in terms of test scores, Stake's "responsive evaluation" tries to employ a more holistic approach.

Semi-structured evaluation. Although many of Stake's criticisms of traditional evaluation methods are certainly to be heeded, it is by no means clear that tests should be abandoned or even demoted in importance. Rather we feel that evaluators must become more cautious in their interpretations of test results and must become more flexible in their use of other approaches to evaluation (e.g., by including the reports of trained classroom observers in evaluations). The need for flexibility and new evaluation methods is especially pressing in research in informal logic. The Illinois Rational Thinking Project is examining several methods for evaluating curriculum materials in critical thinking. We have found that tests alone do not provide all the information we would like, although we still consider them an indispensible part of evaluation. We are beginning to examine other evaluation methods, some of which are indicated below. Whether these techniques will prove useful remains to be seen, but we invite others to experiment with them and hope others interested in informal logic will make additional suggestions.

Many skills that informal logicians wish to teach their students are not amenable to evaluation by means of traditional tests. For example, the application of informal logic skills in conversation and in everyday arguments is an extremely complicated process. By observing human interactions that are more or less structured, one can begin to get a feel for students' abilities in this area. On the more structured side, debates provide a format that might even produce quantitative data if some type of scoring system is employed. Students must both construct and criticize arguments in a debate, so this particular activity is one which teachers of informal logic should consider using in their classes. Scoring procedures need to be developed *by informal logicians,* since what they perceive as good and bad in a debate is different from what the rhetorician sees as good and bad.

Debates, while useful in the evaluation of instruction, are not very realistic forums for the application of logical skills. One problem with them is that they do not allow participants to change their positions when they hear a good argument from an opponent (see Scriven, 1976). Small group discussions might provide a more realistic setting for the application of logical skills in a context likely to be found in everyday life. An interview situation might also provide a good context in which to evaluate the ability of students to construct and criticize arguments. Like debates, discussions and interviews might lend themselves to analysis by means of a scoring system, especially if the topic is one in which certain lines of argument could be expected. However, remembering Stake's criticisms of traditional methods, one should not rely exclusively on a scoring key when evaluating something as open-ended as a discussion or interview. An evaluator must be able to spot unforeseen moves that would indicate that students are employing the skills and concepts that have been taught.

Surveys and questionnaires. While surveys and questionnaires are used to some extent at the present time, they are not being employed as fully as they could be. At the primary and secondary level how a course is perceived by other teachers, parents, administrators, and, especially, the students themselves can be important factors in the success or failure of a course. At the college level, how the course is perceived by students is still a very important factor. Whether students view a course as training in the rational pursuit of truth or as training in sophistry will certainly affect our evaluation of the course. Some attempt is now made to analyze surveys on the basis of traditional test theory models, but it is not at all clear that that model is appropriate. More investigation is needed in this area.

Long-term follow-up. Probably the most neglected approach to evaluation is the long-term follow-up study. While this approach might fit under traditional testing models, this depends on the type of follow-up performed. Unfortunately these kinds of studies are rarely done. This is a rather sorry state of affairs since the effects of most educational programs are meant to be lasting. However, most programs and courses are evaluated at the end of the treatment period and follow-up studies are very expensive and difficult. Lindquist (1951) distinguishes between immediate objectives, those which end-of-course evaluations measure, and ultimate objectives, the attainment of which can perhaps only be evaluated at some time long after the treatment period. We suspect that informal logicians will be especially concerned with ultimate objectives, since informal logic courses are meant to help people reason in everyday situations throughout life. Without long-term follow-up studies, it is difficult to see how one could decide whether ultimate objectives had been attained. The financial and logistical problems which are inherent in long-term studies are obvious, but this fact does not reduce the need for such studies. Rather it counsels us to be well-prepared (and supported) before venturing such a study.

SUMMARY

We have presented a brief overview of the state of testing and evaluation as applied to informal logic.

Currently available general tests in this area were described and criticized. They are the *Watson-Glaser Critical Thinking Appraisal*, the *Cornell Critical Thinking Test, Level X*, the *Cornell Critical Thinking Test, Level Z*, and (if one groups them together) the *Instructional Objectives Exchange Indexes*.

Two standard, generally-desirable characteristics of tests were explained: reliability, the tendency of a test to give the same result when given again in the same circumstances; and validity, the characteristic of measuring (or appraising) what the test is supposed to measure (or appraise). Test-retest, parallel form, and internal consistency methods of estimating reliability were described and criticized, and the danger of using multiple internal-consistency methods for tests of heterogeneous traits was noted. Five common approaches to validity were considered: face, content, construct, predictive, and concurrent. Current test-specialist contempt for face validity was questioned; the notion of content validity was challenged because of its intimate relation to the problematic concept, *universe of behaviors;* construct validity, the idea that a test is valid to the extent that its results fit into a good theory, was explored and found vague, but not uselessly so; and predictive and concurrent validity were deemed to be generally of little use to informal logicians because of the lack of an outside criterion to validate informal logic tests.

We distinguished criterion-referenced testing from norm-referenced testing. In doing so, we suggested a shift in testing-theory vocabulary from "test" to "testing," the reason being that a test developed for one purpose could conceivably be used for the other. Criterion-referenced testing has the purpose of assessing degree of mastery; norm-referenced testing has the purpose of discriminating between and among students and groups. Norm-referenced testing theory is well-developed, though there are problems, including the built-in invitation to develop reliable, invalid tests. Criterion-referenced testing theory is in its infancy, and in particular has problems with its lack of guidelines for determining a level that shall be deemed mastery and with its generally-accompanying concept, *universe of behaviors*. It is not clear whether the recommended random sample from the universe is to be taken of behaviors as dispositions, of behaviors as performances, or of items.

Some procedures for developing one's own informal logic tests were suggested, and various types of evaluation instruments (in addition to the heavily-emphasized multiple-choice tests) were described and recommended.

Experimental designs were considered. We do not recommend the simple pretest-posttest design unless there is a control group. But even if one has a control group, experimental theory calls for the random selection of the subjects for the experimental and control groups *from the population about which we want to draw conclusions*. This is impossible if we want to draw conclusion about next year's classes, for example, so compromises are struck. One compromise is to draw one's initial conclusions only about the group from which one did manage to draw a random sample, and then attempt somehow to infer to the larger group on the basis of its typicality. A second compromise that is often struck is to pick one's experimental and control groups not at random, but so that they are as comparable as we can get them, and then to assume that they are comparable enough, or to use statistical techniques that, it is hoped, compensate for incomparability (this is called a "quasi-experimental design"). There is no perfect resolution of these problems.

As we proceeded in laying out this introductory treatment of testing and evaluating in informal logic, we broached a number of philosophical problems that are embedded in this field. We did not attempt an exhaustive list of such problems, but did allude to the following: what sense can be made of random sampling from a universe of behaviors? What is a "behavior"? What is a true score? What is critical thinking? What is rational thinking? What is informal logic? What is the relationship between test performance and mental traits? What is mastery and in general how can mastery be inferred from test performance? Is it plausible to judge a test to be valid on the ground that it fits into a well-confirmed theory, as is recommended by the construct-validity approach? If so, then what rules and procedures can be followed to make such judgments? What constitutes typicality? Can one specify guidelines for generalizing beyond a population from which a random sample was drawn? If so, what are they? Can one specify guidelines for acceptable alternatives to random sampling? If so, what are they?

We mention these problems partly in order to warn interested informal logicians that the field of testing and evaluation is not out there all ready to provide a neat, clean service to us. But we do so also in the hope that some philosophers will undertake work on these or other evaluation-related problems with the intention of offering theoretical help in this area. In view of informal logicians' practical interests in evaluating informal logic competence, it should be apparent that philosophical work on these problems would be a socially-significant activity. We also feel that such work is intrinsically interesting and philosophically important.

We also hope that other informal logicians will develop various kinds of instruments for evaluating informal logic competence. More are needed, and if *we* do not do it, someone else will—someone who knows even less about it than we do.

REFERENCES

Ahman, J. S., & Glock, M. D. *Evaluating pupil growth.* Boston: Allyn and Bacon, 1958.

Airasian, P. W. Designing summative evaluation studies at the local level. In W. J. Popham (Ed.), *Evaluation in Education.* Berkeley, Calif,: McCutchan, 1974. Discussion of choosing an experimental design under the constraints imposed by typical institutional settings.

American Psychological Association. *Standards for educational and psychological tests.* Washington, D.C.: American Psychological Association, 1974.

Anderson, R. C. How to construct achievement tests to assess comprehension. *Review of Educational Research,* 1972, *42,* 145-170. Contains practical suggestions for writing types of items which are useful in criterion-referenced tests.

Bechtoldt, H. P. Construct validity: A critique. *American Psychologist,* 1959, *14.*

Bloom, B. S., Englehart, M. D., Furst, E. J., Hill, W. H., & Krathwohl, D. R. *Taxonomy of educational objectives, handbook I: Cognitive domain.* New York: David McKay, 1956.

Buros, O. K. (Ed.) *The seventh mental measurements yearbook* (2 vols.). Highland Park, N.J.: Gryphen Press, 1972. Standard sourcebook for psychological tests.

Campbell, D. T., & Fiske, D. W. Convergent and discriminant validation by the multitrait-multimethod matrix. *Psychological Bulletin,* 1959, *56,* 81-105.

Campbell, D. T., & Stanley, J. C. Experimental and quasi-experimental designs for research on teaching. In N. C. Gage (Ed.), *Handbook of research on teaching.* Chicago: Rand McNally, 1963. Classic article on experimental designs for educational research.

Cronbach, L. J. Validation of education measures. In P. H. DuBois (Ed.), *Proceedings of the 1969 invitational conference on testing problems.* Princeton, N.J.: Educational Testing Service, 1969. Preliminary version of Cronbach, 1971.

Cronbach, L. J. Test validation. In R. L. Thorndike (Ed.), *Educational measurement.* Washington, D.C.: American Council on Education, 1971. A seminal paper on test validity. Some parts are technical, but it is recommended reading, nonetheless.

Cronbach, L. J., & Furby, L. How we should measure "change"—or should we? *Psychological Bulletin,* 1970, *74,* 68-80. Arguments against the use of gain scores and the pretest-posttest control group experimental design.

Cureton, E. E. Reliability and validity: basic assumptions and experimental designs. *Educational and Psychological Measurements.* 1965, 25, 327-346.

Ennis, R. H. An appraisal of the Watson-Glaser critical thinking appraisal. *Journal of Educational Research,* 1958, 52, 155-158. Covers earlier versions of the Watson-Glaser test (Forms Am and Bm).

Ennis, R. H. Assumption-finding. In B. O. Smith & R. H. Ennis (Eds.), *Language and concepts in education.* Chicago: Rand McNally, 1961.

Ennis, R. H. A concept of critical thinking. Harvard Educational Review, 1962, 32, 81-111. With a few minor amendments, this notion of critical thinking is the basis for the Cornell Critical Thinking Tests.

Ennis, R. H. Operational definitions. *American Educational Research Journal,* 1964, 1, 183-201.

Ennis, R. H., Gardiner, W. L., Morrow, R., Paulus, D., & Ringel, L. *The Cornell Class Reasoning Test.* Urbana, Ill.: Illinois Critical Thinking Project, 1964.

Ennis, R. H., Gardiner, W. L., Morrow, R., Paulus, D., Ringel, L., & Guzzetta, J. *The Cornell Conditional Reasoning Test.* Urbana, Ill.: Illinois Critical Thinking Project, 1964.

Ennis, R. H., & Millman, J. *Manual for Cornell Critical Thinking Test, Level X and Cornell Critical Thinking Test, Level Z.* Urbana, Ill.: Illinois Critical Thinking Project, 1971(a).

Ennis, R. H., & Millman, J. *The Cornell Critical Thinking Test, Level X.* Urbana, Ill.: Illinois Critical Thinking Project, 1971(b).

Ennis, R. H., & Millman, J. *The Cornell Critical Thinking Test, Level Z.* Urbana, Ill.: Illinois Critical Thinking Project, 1971(c).

Ennis, R. H., & Paulus, D. *Critical thinking readiness in grades 1-12* (phase I: deductive logic in adolescence). Ithaca, N.Y.: Cornell University, 1965. (ERIC Document Reproduction Service No. ED 003 818).

Glaser, R., & Klaus, D. J. Proficiency measurement: Assessing human performance. In R. M. Gagne (Ed.), *Psychological principles in systems development.* New York: Holt, Rinehart, and Winston, 1962.

Glass, G. V. Standards and criteria. *Journal of Educational Measurement,* 1978, 15, 237-261.

Guilford, J. P., & Hertzka, A. F. *Logical Reasoning* (test). Orange, Calif.: Sheridan Psychological Services, 1955.

Hambleton, R. K., Swaminathan, H., Algina, J., & Coulson, D. B. Criterion-referenced testing and measurement: a review of technical issues and developments. *Review of Educational Research,* 1978, 48, 1-47. A review of the state-of-the-art in criterion-referenced testing. Some technical sections.

Haney, W. V. *The uncritical inference test.* Wilmette, Ill.: William V. Haney Associates, 1975.

Hempel, C. G. A logical appraisal of operationism. In P. G. Frank (Ed.), *The validation of scientific theories.* New York: Collier, 1961.

Hempel, C. G. *Aspects of scientific explanation.* New York: Free Press, 1965.

Hempel, C. G. *Philosophy of the natural sciences.* Englewood Cliffs, N.J.: Prentice-Hall, 1966.

Instructional Objectives Exchange. *Judgment: deductive logic and assump-*

tion recognition, grades 7-12. Los Angeles: Instructional Objectives Exchange, 1971.

Kelley, T. L. *Interpretation of Educational Measures.* Yonkers, N.Y.: World Book Co., 1927.

Kerlinger, F. N. *Foundations of behavioral research.* New York: Holt, Rinehart, and Winston, 1964. In-depth discussions of experimental designs.

Lindquist, E. F. Some preliminary considerations in objective test construction. In E. F. Lindquist (Ed.), *Educational measurement.* Washington, D.C.: American Council on Education, 1951.

Lord, F. M., & Novick, M. R. *Statistical theories of mental test scores.* Reading, Mass.: Addison-Wesley, 1968. The definitive work on norm-referenced test theory.

Messick, S. The standard problem: meaning and values in measurement and evaluation. *American Psychologist,* 1975, *30,* 955-966.

Petrie, H. Against objective tests: a note on the epistemology underlying current testing dogma. In Mark Ozer (Ed.), *Toward the more human use of human beings: Issues in the application of cybernetics to assessment of children* (forthcoming).

Popham, W. J. (Ed.). *Criterion-referenced measurement.* Englewood Cliffs, N.J.: Educational Technology Publications, 1971. Good introduction to criterion-referenced testing.

Sax, G. The use of standardized tests in evaluation. In W. J. Popham (Ed.), *Evaluation in education.* Berkeley, Calif.: McCutchan, 1974. Contains a comparison of criterion-referenced and norm-referenced tests.

Scriven, M. *Reasoning.* New York: McGraw-Hill, 1976.

Sell, D. E. *Evaluation aptitude test-manual.* Munster, Ind.: Psychometric Affiliates, 1952.

Smith, R. A. *Regaining educational leadership: Critical essays on PBTE/CBTE, behavioral objectives, and accountability.* New York: John Wiley, 1975.

Stake, R. E. The countenance of educational evaluation. *Teacher's College Record,* 1967, *68,* 523-540.

Stake, R. E. To evaluate an arts program. *Journal of Aesthetic Education,* 1976, *10,* 115-133.

Stake, R. E., & Hoke, G. A. Movement and dance in a downstate district. *The National Elementary Principal,* 1976, *55.*

Stanley, J. C. Reliability. In R. L. Thorndike (Ed.), *Educational measurement* (2nd ed.). Washington, D.C.: American Council on Education, 1971. A thorough treatment of the topic. Somewhat technical.

Stewart, B. *Testing for critical thinking: A review of the resources.* Urbana, Ill.: Illinois Critical Thinking Project, 1979.

Strawson, P. F. *Introduction to logical theory.* London: Methuen & Co., 1952.

Watson, G., & Glaser, E. M. *Manual for Watson-Glaser critical thinking appraisal.* New York: Harcourt, Brace, & World, 1964(a).

Watson, G., & Glaser, E. M. *Watson-Glaser critical thinking appraisal, form Ym.* New York: Harcourt, Brace, & World, 1964(b).

Watson, G., & Glaser, E. M. *Watson-Glaser critical thinking appraisal, form Zm.* New York: Harcourt, Brace, & World, 1964(c).

Wesman, A. G. Writing the test item. In R. L. Thorndike (Ed.), *Educational measurement.* Washington, D.C.: American Council on Education, 1971.

Wiersma, W. *Research methods in education* (2nd ed.). Itasca, Ill.: F. E. Peacock, 1975.

Winer, B. J. *Statistical principles in experimental design* (2nd ed.). New York: McGraw-Hill, 1971.

ADDITIONAL READINGS

Freedman, D., Pisani, R., & Purves, R. *Statistics.* New York: Norton, 1978. Introductory text.

Gage, N. L. (Ed.) *Handbook of research on teaching.* Chicago: Rand McNally, 1963.

Glaser, R., & Nitko, A. J. Measurement in learning and instruction. In R. L. Thorndike (Ed.), *Educational measurement.* Washington, D.C.: American Council on Education, 1971. Contains a discussion of some uses of criterion-referenced tests.

Glass, G., & Stanley, J. C. *Statistical methods in education and psychology.* Englewood Cliffs, N.J.: Prentice-Hall, 1970. Widely used text, moderate level of difficulty.

Millman, J. Criterion-referenced measurement: Current applications. In W. J. Popham (Ed.), *Evaluation in education.* Berkeley, Calif.: McCutchan, 1974. A thorough introduction to criterion-referenced testing.

Scriven, M. The methodology of evaluation. In R. W. Taylor, R. M. Gagne, & M. Scriven, *Perspectives in curriculum evaluation.* Chicago: Rand McNally, 1967. Scriven here introduces an important distinction between formative and summative evaluation.

Footnotes

[1]We would like to thank Robert Linn, Stephen Norris, Denis Phillips and Bruce Stewart for their comments on and criticism of a draft of this paper.

PERORATION

THE PHILOSOPHICAL AND PRAGMATIC SIGNIFICANCE OF INFORMAL LOGIC

Michael Scriven
University of San Francisco

This paper consists of a series of notes falling under a very extended conception of this very extended title. I think this most important occasion calls for the widest possible range of reflection on our topic and on our experiences, and maybe even a little rhetoric, because the beginning of a new era is the time for considering all possibilities as we plan our time and our activities in the years to come—and the time for a little inspiration.

1. Let me start with some thoughts about the philosophical significance of informal logic.

To begin with, the emergence of informal logic marks the end of the reign of formal logic. Not by any means the end of the *subject*, just its relegation to its proper place in the academic zoo, somewhere over there just north of mathematics and west of computer science, and far away from the children's part of the zoo. It's not *good* for children to see too much of the monsters there; it warps their little minds, gives them dread diseases like Meinong's syndrome and quinea and the kripkes. They grow up into poor little perverts who—in the case of Tarski psychosis—mutter things like " 'p is true' if and only if p," then smile beatifically. Or they go around chanting, "A false proposition implies any proposition, yes it does, yes it does—and *any* proposition implies a true one, so it does too, so it does too." They exhibit curious semantic allergies; for example when shown patterns of symbols like this:

Most A's are B's
Most A's are C's
Some B's are C's

they shake their heads convulsively, muttering, "No, No, it's invalid, invalid." But as Sir William Hamilton pointed out a very long time ago, "most" is a quantifier like "all" and "some" and "none" and quite obviously the above inference is the basic one it legitimates. (The example is from Geach, **Reason and Argument**.)

For a long time people thought all the beasts in the logic cages came from the real world, perhaps from a subterranean part of it where everything ideal resided, though it wasn't easily seen. Now that we realize those weird creatures, like the ones called "Paradoxes of Implication," are mostly just fakes, we should let them fade away in peace. It releases a great deal of valuable room.

In short, logic has—with the emergence of informal logic—been called back to its proper task, away from the pathology. It may or may not be in time to save philosophy. The Wittgenstein revolution in philosophy provided an opportunity for salvation, but—generally speaking—the opportunity was missed. The great philosopher of ordinary logic analysis was all too quickly converted into a cult figure whose followers wallowed in ordinary logic as if it was the Holy Ganges and not just a spring of fresh water, forgetting most of the tasks of philosophy or simply dealing with logical idealizations of them not as a means to the ends of philosophy but as ends in themselves, thus renouncing Wittgenstein's own advice. This is one of the pseudo-professionalism traps in philosophy, like the trap of making philosophy into a study of what Socrates said instead of what Socrates studied. Russell didn't drown in math logic, he went through it as an adolescent phase. Leibniz's hope for the *calculus ratiocinator* was not that it would replace chess as a fun game, but that it would solve *philosophical* problems. And so on.

In short, this movement is not just a way to pablumize baby logic, it is the last hope—or at least the latest hope—to save logic from lunacy.

The forces against it—as against most innovations—are incredibly powerful. They include natural laziness, professionals' ego-involvement with their past work, straight disbelief, the sense of incompetence in handling the new material and approach, lack of interest (though this is very much a matter of fickle fashion), threat to turf and enrollment, ignorance, moral indignation at "lowered standards," the need to teach jargon in order to be sure you've taught something, professional jealousy, and the tightening of resources. These unattractive attitudes and conditions are to be found not only in many of our philosophical colleagues but in the other departments that have grabbed a piece of the action, in rhetoric, English, journalism, communications, general studies, composition, and so on.

I don't know if we'll make it. I am sure that right is on our side but then of course I have some difficulty in defining "right," by traditional standards of definition.

Is it not indicative of the absurdity of standard philosophy that the informal logic movement (I suppose it is one) should find much of its opposition amongst the ordinary-language establishment? They came to adopt formal logic, perhaps because they were taught it, without so to speak looking at it, without noticing that it is the enemy. Thus do social forces control even the discipline of reason itself.

Why all this enthusiasm about informal logic? The reason implicit in the preceding is simply that it represents a turning back to a proper task of philosophy in general and of logic in particular, namely the study of argument. It is as if philosophy of science had for years studied no real sciences, but only certain formal reconstructions of them by philosophers of science. The latter kind of study is *essentially* secondary, a sideline, but I have never been persuaded that formal logic has ever contributed anything significant to the understanding of any problem that it did not create. It's supposed to solve *other* people's problems, not create its own.

Now the analogy to philosophy of science brings us to a second crucial reason for welcoming the new logic. For philosophy of science *has* been substantially guilty of the hypothetical crime I described. And logic connived at the crime, compounded it, sometimes created it. Explanation-theory became a mess at least partly because Hempel and others thought material implication was a fair formalization of inference. If you recall his reasons for arguing that the explanation of general laws *could not* be like that of particular events, it was that one could then generate paradoxical consequences via the usual flaws in the horseshoe. No one ever picked this up—yet if they had we would soon have seen what else was wrong with the inferential model of explanation. For nothing could be more absurd than the idea that scientific explanation of individual events was logically something different from scientific explanation of patterns in events, especially given that our language for singular events presupposes regularities.

That is, formal logic not only dirtied its own nest, it dirtied philosophy of science—and ethics and philosophy of math and so on. So a switch to common sense standards can save other parts of philosophy. That's a great part of the philosophical importance of informal logic.

But we can go deeper than this. We can look at the *logical foundations* of informal logic, the concepts and distinctions and relationships that are necessary in order to make sense out of the procedures for the criticism and construction of arguments in science and everyday life. When we do this I believe we find some extremely important concepts emerging which themselves lead us on to revolutions in other branches of philosophy. A most important set of examples of this is the way in which we are forced to reconsider a certain family of old chestnuts, the "fallacy of psychologism," "circularity," the "genetic fallacy," the deductive/inductive distinction, the "naturalistic fallacy," the "context of discovery/context of justification" distinction and the cause/reason distinction. Naturally, reconsideration of the *traditional* fallacies in logic, as is being done by many of our colleagues in informal logic, is a valuable enterprise indeed—for logic. Reconsideration of these other (supposed) fallacies is valuable for the whole of philosophy. I wish I had time here to do even one of these in detail—perhaps there will be a chance in the discussion period. Let me just try to make plausible in one case the claim that the consequences would be important, taking the case of the so-called fallacy of psychologism, of confusing logical issues with psychological ones.

It has long been obvious that the concept of explanation, in science or elsewhere, is neither syntactic nor semantic but essentially pragmatic (to use a traditional set of terms that are themselves somewhat contaminated by formalism). To be more precise, explanation cannot be analyzed without reference to the concept of understanding and thus ultimately to the extent to which particular individuals understand particular matters, i.e., their state of mind, i.e., psychologism. I have argued for many years that this is true of each of the main concepts in the structural grammar of science—of prediction, observation, probability, law, approximation, classification and so on. In short, philosophy of science and hence science are far more crucially psychologistic—or shall we say subjective (in the sense of person-related) than has been believed. And I find the same point to hold throughout informal logic—the notion of proof, of assumption, of fallacy, of truth, of reason, of argument, of persuasion, of consideration, of inference, of probability—all of them are at the least dualistic notions whose mentalistic face has been veiled for too long. Possibly the best new primitive term to use is information, the organization of which is what generates understanding. For example, we have to realize that tautologies are sometimes extremely informative even though they may lack empirical content. Nor are they "merely" informative about language; they are informative about the world, they clarify our thinking about it by listing the options or unpacking the complexes in it. (The law of effect and indeed the laws of motion are good examples.)

So informal logic may provide the extra load that breaks the back of the old neo-positivist epistemology as well as logic. I suspect this will turn out to be true.

There are other crucial implications for the rest of philosophy in informal logic. No serious study of practical evaluation can fail to reveal at least three legitimate ways to end-run the fact/value distinction and the naturalistic fallacy. That mortally wounds many ethical theories and criticisms of others. The philosophy of history and the genetic "fallacy" are intimately entwined. And so on.

Do not let yourselves forget that huge gains for formal and philosophical logic are possible. One thinks here of massive extensions of the Anderson and Belnap work, of the need for unpacking the continuum from transformation to proof, the reinterpretation of the incompleteness results, the overdue need to look at computer programming languages, especially PASCAL and FORTH to learn more about alternative approaches to axiomitization (and to develop better programming languages, since even these—best—are logically confused); and so on.

In one way, I'm saying the philosophical importance of informal logic is that it generates a new perspective on philosophy. John Woods was saying last night how he had noticed ties between some of his work and Kripke's semantics. The proper perspective, I think, is that if we'd been doing logic properly we'd have a better philosophy than Kripke's by now and *he* could work out how to tie to us. It was obvious even before Toulmin stated it, decades before Kripke, that hierarchies of possible worlds (of possibility spaces) are essential to handle the range of implication relationships. The informalists just never worked it up in detail and of course Kripke hasn't either. There are more of us now—perhaps enough to save logic from the schoolmen's fate.

And then there are the implications for and from the social sciences—the failure of the positivist model there and thus of formalization in that sense, leaving open a wide range of possible alternative ways to rigorize and reconstruct—and, it must be added, to improve and extend. The alternative to deontic logic is not situational ethics; the alternative to axiomatized social science is not anecdotes but disciplined geography and history and behavioral sciences with their own reconstructed logic of sensible inference, not the straitjacket of standard symbolic logic—or its multi-valued mutations. Dozens of scientific specialties need a new logic. It so happens that I've been writing on clinical inference lately and skimming the literature on person perception—in both cases, the simplest attempts to express the models are crippled because no one has worked hard on ways to express these inferences. Reviewing my notes for a paper on cognitive psychology, I see the same problem. And in the field which I now call home most days of the week, the new discipline of evaluation, it is not an exaggeration to say that one's capacity to cope with it conceptually wholly depends on understanding the difference between statistical inference, theoretical inference and prima facie inference. And where would one send an inquirer for answers on that? Where, on the other hand, would one find a better group of people to generate answers than here?

We have slighted many other areas in this swift survey, perhaps most seriously the law. The whole mighty effort of the New Rhetoric is an informal logic battlefront, badly in need of reinforcements, full of potential plunder, fat with fascination. There are many others, from quantum field theory to taxonomy.

In sum, informal logic can serve as the fountainhead for a new flood of intellectual revolutions, not only in logic and philosophy, but across the whole landscape of the mind.

2. I'll discuss the *practical* significance of informal logic under three headings: the doctrine, the pedagogy, and the movement.

While there are of course practical consequences of the informal logic approach to the social sciences which I've already covered, in this part of the paper I'll concern myself solely with examples of *direct* practical applications and implications.

As a first example, let's take the practical procedure of so-called "psychological" testing; in particular, achievement tests. Here the skills that are developed in teachers and hopefully in students of informal logic classes are relevant to a startling degree. People, including professional test designers, have lost touch with the ordinary sense of words and developed hidden meanings to which the testees are not all privy, with disastrous results for the validity of the tests. There are standard college entrance tests on the market where students are requested to say what the causes of the Civil War were, and where the *scoring key* reveals that giving the causes of the Civil War comprehensively, clearly and succinctly could not possibly get a high grade. A secret agenda is involved, according to which bright students are taught by their history teachers to free-associate to such questions. The sample good answers are in fact full of material that is specifically *irrelevant* to the question. How can this happen? It happens because history teachers from "the better schools" are the source of many of the questions and scoring keys that are used in these tests, and thus bring in the hidden standards. This is politically important for the sales of the tests, and for their credibility, which is their *apparent* validity. Their true validity, however, depends on the *logical* question of the relevance of these answers to the questions. This kind of consideration is usually dismissed by testing types as "mere face-validity." It is not so easily dismissed; no empirical data about the high correlations between independently identified high achievement levels in students and high scores on this kind of test can make it valid in the absence of the logical connection. But this is a typical *informal* logic point and one that has not been incorporated into the training of test-makers, because it's a "soft" consideration not a statistical or experimental one.

One could multiply examples indefinitely, but I'll just mention a couple more, and in doing so spell out some of the implications for pedagogy. The practical implications of all these examples are extremely important because of the extent to which the new push for basic education rests upon a capacity to identify basic skills. Since soft logic skills are not commonly recognized as a necessary part of the basics curriculum, whereas they are in fact the glue that connects knowledge of facts with any applications of that knowledge, the very concept of that curriculum is in trouble. Pushing back to the most primitive soft logic skill—understanding prose, which involves identifying its first level implications—we find that precis and transliteration, which are the key relevant test performances, are almost never found in the standardized test battery. This is, of course, because they are not multiple-choice tasks and hence are quite difficult to grade reliably. And there *are* multiple-choice tests which correlate quite highly with them. But that's not good enough for a *good* teacher; we should insist on having students perform *these* tasks and we should help them to improve *those* performances because quite a large slice of the variance in the scores is potentially under the teacher's control. What one is picking up with the correlated multiple-choice tests is mostly the part that is *not* under the teacher's control, i.e., IQ plus early-environment effects. Not surprisingly, if we use *these* tests as tests of improvement, we will be disappointed in our efficacy. Once more, sloppy thinking about thinking gets us into trouble.

Now, to abandon somewhat irrelevant "objective" tests for very relevant but highly subjective tests, i.e., tests which are scored very differently by different teachers, or even by the same teacher at different times, is no bargain. Here we have to turn our attention towards another underground literature besides our own, the series of mostly fugitive papers by Paul Diederich on how one can get one's own (and others') reliability as an essay grader up to a high level. I've read these once, in the ETS library in Princeton, but they deserve anthologization, commentary and great respect. I hope the informal logic movement may be instrumental in getting more attention paid to this aspect of the gap between what we do as teachers and what students need. It was obvious that teaching math logic was a long way from their needs; it's less obvious that teaching soft logic is just as far away if the validity and the reliability of our tests (and our comments on them) is low.

English composition itself, both its testing and its teaching, which one might think of as a prelogical skill, is unfortunately infected with illogic. Look at the test items for the National Assessment of Educational Progress writing tests and you'll find an analog of the error discussed above for a history test. One assignment is to describe what's happening in a picture that shows a stag swimming a river in a blazing forest. The *bad* "model answers" do just that; the *good* ones go off into wild fantasies about how the fire might have started. A little experiment we have recently done in the course of a 3-year Carnegie-financed evaluation of the Bay Area Writing Project turned up a related problem—the incapacity of experienced judges from the composition field to separate logic from style considerations, in particular their tendency to overgrade stylish nonsense and undergrade crude but basically correct criticism. We will not get far in "basics" if we don't know how to make these distinctions—but the whole tradition in the "humanistic" areas, pre-college, has drifted towards the adulation of poetic maunderings rather than inelegant common-sense. The impact on teaching practical skills is devastating, because we are not only failing to appreciate it when it does appear but encouraging the brighter students to ignore it, indeed *to turn away from it*, as we guide their efforts to improve.

Incidentally, the Bay Area Writing Project is a good project: it would be still better if it transcended the standards of the field and went after reasoning skills per se—merely as one of several desiderata, of course. Perhaps we need comparable projects in soft logic—but I'd much rather see the two combined.

If you want further examples along these lines, you might want to look at the LSAT (Law School Admissions Test), one of the better examples of well-reasoned testing but not without its own little biases; or the Graduate Record Exam.

It's very important in all this to be conscious of the existence of a better tradition in Canada, one which deserves and needs support. This tradition comes in part from the British influence, where the high school English curriculum has always had a very decent soft-logic component—one thinks of R. H. Thouless' *Thinking Straight* and many other good little texts. And there are provinces where the provincial department of education has lent some strong support to these efforts—I think of Alberta's stress on moral reasoning, in part a tribute to Simon's work at Calgary.

We've talked about the humanistic side; equally fascinating is the failure to see the crucial importance of evaluation as a key reasoning skill in *science*, and the failure to try for teaching *generalizable* reasoning skills in science courses (something I did try for, not very well, in **Applied Logic: an Introduction to Scientific Method**, 1965).

My conclusion from all these examples is that the testing and teaching process could stand a thorough overhaul at the hands of informal logicians. We can't settle for text-writing; we've got to get into the rest of the teaching process. (Here we are fortunately able to point with great pride to one of our colleagues in The Movement, Bob Ennis, who long ago saw some of these problems and proceeded to do something—in fact, a great deal—about them. I feel we are very fortunate to have his major contribution in this volume.) These further changes will not be easy to introduce; they may require multi-media presentation, or multi-color printing, or the use of psychodrama. I am already having trouble with my publisher over the implications of this for the 2nd edition of **Reasoning**, which in content, format, and approach—I now think—will have to be one or two quantum jumps more unconventional than the first. I will probably have to publish it myself, as I did the first edition of that and this. The usual pubishers, commercial or academic, are establishment-controlled.

Before going on to pick up another type of example that bears on pedagogy, I want to mention a last category of practical payoff from informal logic. I have in mind the need for enormously improved analytical capability with respect to contemporary problems. My sense is that if we spent half the time on soft logic analyses of contemporary problems that we do on national history in the schools, we would double the chances of there *being* a national history this time next century. The curriculum time might as well come out of national history anyway because that already has an indefensibly large slice of the action, and what's left would be much improved if it could be approached with some capacity for soft logical analysis. (It's also very doubtful whether the lost 50% of seat hours would show up as even a lost 5% of learning retained six years later.)

Now, why should we get into this expenditure of student/teacher time and effort? Simply because our political representatives at every level display a lack of capacity for elementary problem analysis that we in this group feel sure they could have been taught; and because the costs to all of us from this are awful now with worse to come. The kind of history we teach helps little or not at all; the kind of science we teach does no better. Look at the discussions of inflation; nearly all focus on contradictory technical solutions from economics which are well-tested and palpable failures. The solutions —and there are probably several that would work—have to be *system* solutions and have to address mainly the psychological link in the vicious circle (not with mere rhetoric) and have to be much more radical than nudging the rediscount rate. So much is obvious from looking at a few examples of rampant inflation and its outcome; but where did you see a serious analysis of what worked in Switzerland a few years ago and why it hasn't worked here, of what happened in Germany and Bolivia and France last time? Perhaps in an economics journal? Why not in newspapers which by now have reached a very high level in purveying medical, horticultural, psychological and meteorological advice. Good old Mill's Methods to work fine on economics; what *doesn't* work is the "science" of economics, and what we haven't done is to get to the point of using an extended common-sense approach there. We're still awed by the scientific mystique in an area where it has never been earned (plus the undeniable complexity of the issue). Even the Nobel Prize committee was conned into adding economics as the only social science to the list of prizes; conned by the formalism, ignoring the common-sense standards.

Read the papers looking for the best arguments of strikers, especially striking public servants; you can almost never find them—a violation of the First Law of Argument Assessment. Look for systematic tracings of the cost of increasing population size, the fuel that runs inflation and unemployment, destroys environments (and the quality of life in other ways) and drives up taxes, starvation and crime rates. Are its benefits worth these and other costs? Who tries to set these out in tabular form, regularly updated, so the citizens can attach their own weights and come to a conclusion? The media are simply bad at handling complex problems, which they reduce to either a series of flashes about who's on which side, perhaps with one-line quotes, or a historical background story; neither is sufficiently analytic to be adequate. Basically, there is little hope for improvement here until *either* most journalism majors *or* a critical mass of opinion-makers get the sense of power and competence with complex arguments that good soft logic devices provide. (I like to think of soft logic as the intellectual equivalent of low-technology solutions, à la Schumacher.) Look carefully at the logic of the pro- and anti-abortion arguments; usually they are textbook examples of fallacies in practical reasoning. Consider the rules governing conflict of interest for public servants (or academic tenure); they are textbook examples of generalizations to which long chains of counter-examples can be formulated. I need hardly mention the usual discussions of marijuana or cocaine or heroin, affirmative action, the death penalty or the speed limit, since we've already tapped those veins in our own courses so often. My point is simple. Contrary to the popular diagnosis that these are examples of how small a part reason plays in human life, that they tell us something important about the limits on human capacity to be rational, I take the much simpler view that they just show how little skill in reasoning we teach. Reasoning *takes* teaching; and reasoning about practical problems takes teaching reasoning *using* tough, practical, controversial problems as examples. Learning very rarely transfers unless it is designed to do so, and helped along; but *we* have set our primary and secondary schools up so that with few exceptions it is professionally impossible to survive as a teacher or administrator if you insist on the explicit use of tough practical controversial examples. That must be changed if we are to survive; meanwhile we must capitalize on the residual freedom of discussion in the tertiary educational system (i.e., excluding many—not all—junior and community and sectarian colleges), in order to fill in the crucial gaps in the general education of our future leaders.

The importance of this is so great that it could only be properly handled by making such courses a required part of every degree program. Making our courses good enough to do the job is our first job; the second one, a Movement priority, is to get them into every student's curriculum. That is the harder of the two jobs, but the effort to achieve it will probably quadruple our enrollment anyway, so there is no excuse for despair or sloth!

Training the layperson in soft logic is crucially important, indeed, but we should not leave this area without at least some reference to the practical payoffs for many professionals and for us as their customers. A striking—and depressing—example is provided by the legal profession. Even the best law schools are very weak in teaching argument analysis; they expect the student to extract it from case studies without guidance. Of course, the very best students do just that, but why not assist the process? The "average to good" (by the standards of the field) attorney is, in my extensive experience, really weak; and careful analysis of the lines of argument of the high-fee stars shows them to be a long way from comprehensive. The field has spawned an interesting movement—the so-called New Rhetoric—aimed directly at a semi-soft analysis of legal reasoning. Unfortunately, the ideas so far generated are not of very high quality—but I think this is a prime area for informal logicians to develop, just as philosophers of law have substantially upgraded traditional jurisprudential theory.

Another crucial area is computer science, in particular programming. It is clear that programming languages flagrantly violate soft logical principles and are twice as hard to learn or use for this reason. While this is most obvious in languages like BASIC, where the fine hand of John Kemeny, our formal logic colleague, can be seen (at least between the lines), the new wave languages like PASCAL and FORTH are still wrong in just the way that the propositional calculus is wrong as a model of argument, i.e. they arbitrarily define where they should investigate and describe.

Turning next to an example which is aimed purely at the pedagogical dimension of significance, by contrast with the many hybrid examples we have been considering, I want to ask you all to start looking at educational research, especially research on teaching, with more care than you normally do with respect to material that isn't directly in your field. I have three reasons in mind. First, soft logic exists because we think hard logic didn't do the key *educational* job for logic; that's a functional, pragmatic kind of reason, and the same kind of reasoning must force you into asking the question whether the *way* you teach and not just the content shouldn't be reformed. I think it should: I think, for example, that having students roleplay (and then discuss) bitter arguments and conflict management strategies in class is essential to teaching practical reasoning skills *including the analytical ones*, and it is scarcely part of the standard pedagogical repertoire of most philosophers. (It may be done in some rhetoric/speech/communications departments or by debate coaches, but when I speak of soft logic, I don't intend to refer to loud non-logic or even to soft semi-logic.)

The second reason for looking at research on pedagogy is because it often omits consideration of reasoning *as* pedagogy. Most teaching research (so-called "style" research) concerns the differences between the, e.g., Socratic and didactic mode, or between the use and absence of eye-contact, individualization of assignments, frequent quizzes, etc. (The differences, by the way, usually turn out to be negligible.) Curiously enough, there's very little research on the effectiveness of reasoning itself, by contrast with other forms of systematic presentation. We do know that even highly systematic but unreasoned presentation of moral education material has no detectable effect on moral behavior. We know that reasoned moral presentations of certain kinds can have a large effect (Kohlberg). We do know that highly organized material sometimes *but not consistently* increases student learning compared with disorganized material (text or talk); and the reason it doesn't have the expected effect to the extent or with the consistency most people expected, I speculate, is that there are pedagogical advantages to disorganized presentations in that they *involve* the listener in the *task* of making sense of the material, a task which increases mastery. This (negative) discovery in itself must give us pause. We of all people must avoid the "academic fallacy" in education, the fallacy of the New Math, the fallacy of identifying logically optimal structuring with the pedagogically optimal presentation. So we need to help our educational research colleagues in two ways: to help them see the difference between the organization exhibited by reasoned sequences and other forms of organization; and to help them look for the differences between rational argument and best-teaching procedures. One may find, for example, that in certain areas like practical or applied ethics, reasoning far outperforms other structurings of material; but also that in teaching reasoning, giving bad arguments to be transcended may outperform giving good arguments for emulation (the fallacies approach vs. the ideal language approach?). In any case, large slices of educational research suffer acutely from the failure to be sensitive to the difference between good reasons and poor reasons, partly because making that distinction involves a value judgment, something which empirical researchers like to avoid, and partly because they are often not very good at it since untrained in the relevant soft logic.

Which brings us to the third reason for looking at educational research, the need for us to make further progress with the conceptual foundations of our own discipline, at the point where it impinges on theirs. Thanks particularly to our colleague, Tony Flew (one of us by virtue of his soft logic text as well as the work to be mentioned now), we have a very nice small package of articles in the philosophy of education literature on the distinction between education and propaganda. We need to flesh this out with substantial further thought about the difference between rhetoric, debate skills, and logic. John Wisdom once said, "Logic is proof, proof persuasion, and philosophy logic played with especially elastic equations." This was a scandalously subjective suggestion at the time it was made, but it's not far from a slogan for the soft logic sect. Many of us in soft logic (elastic logic?) must have pondered the lines between presentational skill and arguing skill and reasoning skill. Laying out (structuring) an argument perspicaciously is such a creative process, so intimately connected with graphics and literary skills, that it seems a long way from the critical-analytic-"uncovering" process we think of as paradigmatically logical; and one can't help wondering whether we aren't still suffering from the burden of a false dichotomy (if you'll pardon the expression). The origins of rhetoric and its medieval development are supremely respectable, but we respect little of it today. Why? The analogy with casuistry and sophistry is interesting—those terms of intellectual respect which also became terms of disapprobation, unfairly perhaps. Our new understanding of the history of fallacies should generate new sympathy for the possibility of gold in these mines that most philosophers have long regarded as lacking anything worth serious philosophical respect.

There's a neighboring area which also deserves more attention. As we look at the phenomena of the new cults, from the Patty Hearst "conversion" to the Jonestown pseudo-suicides, we must do it with a real sense of looking behind the scenes of the standard learning process, at the pathology from which one comes to a theory of normal learning. Was Hearst brainwashed or was she rationally persuaded and hence responsible; or irrationally persuaded but subsequently reflective to the point where she inherited responsibility? Our dichotomies need refinement, our understanding needs improvement, and our teaching needs to reflect the results.

Such investigations would not only be valuable for educational researchers doing research on teaching methods, it would also help those of us working on the evaluation of teaching. For presentational skills are somehow connected with the skills of the actor, while also involving the disciplined constraint of the subject-matter. The "Dr. Fox effect" research has made clear that even a sophisticated audience can be conned by a professional actor making a presentation lacking all content. While one must avoid the over-reaction of eschewing all presentational sophistication, the evaluator of teaching (and the evaluated teacher) has to be pretty careful not to use a dowsing rod that is picking up hokum instead of holy water. Is the student rating form subject to this weakness? Or is it rightly reflecting charisma because charisma is a help to good teaching? We need more work on these forms and such work presupposes a better conceptual analysis of good teaching as opposed to enjoyed teaching, as opposed to apparently good teaching, etc., than we have so far done. And I think informal logicians have to clarify the difference between good arguments and persuasive arguments and apparently good arguments before we can distinguish good teaching from apparently good teaching. And I know we're in trouble if we let the formalists get their sticky fingers on *that* job!

So the pedagogical area of praxis is one on which informal logic *can* have great impact and for which it *does* have great implications. One might say that soft logic is nearer to learning than philosophy has been since Socrates and the sophists; but we need to close the gap still further. This entails, I suggest, closer study of recent developments in cognitive psychology, especially the strongly teaching-oriented work on problem-solving algorithms (some of which can be found in a recent anthology from a Carnegie-Mellon conference—*Problem-Solving and Education,* edited by Tunia & Reif), but also, for example, the work on advance organizers by Ausubel and others. Eventually, as I see it, the goal of soft logic is internalizing the skills of reasoning and eliminating the need for complex apparatus; the analog of that supreme example of the alternative to high-technology solutions, the Australian aborigines' almost complete internalization of their extremely sophisticated survival skills.

Finally, let us turn to the practical aspects of informal logic as a movement, or even a Movement. When George Miller gave his famous presidential address to the American Psychological Association a few years ago, on 'giving psychology back to the people,' he reflected part of the stirrings that have surfaced in philosophy with the revolution in the introductory courses, both in general philosophy and in logic, and in events like the "philosophy Woodstock" of Summer 1979 (the official title was "Philosophy, Where Are You?"; see *New Times* October 1979 for a write-up). If we are to keep a semblance of intellectual order in this movement—and the problems with doing that in the women's movement and the parapsychology explosion are quite evident—we'll need to do a little organizing. The question is whether we can find the time and talent to do that, and it's clear that we are much in the debt of Tony Blair and Ralph Johnson for getting us started on the right road. What else do we need? The newsletter is crucial, indeed. But it's going to take more and wider continued support than—as I write a year later—it has so far generated. Basically that means advertising and promoting and exhibiting. In the parallel case of getting the evaluation field going, we have found that carefully chosen mailing lists and space ads are crucial instruments, but a small financial base and good editors are just as crucial. Our next step with the soft logic shuffle may have to be a foundation grant (or more Canada Council support). A booth at the APA meetings would sell subscriptions and get the news around, but it won't generate enough revenue to hire help; that's the kind of situation where a substantial number of volunteers are needed.

Connected with all this PR apparatus, of course, is the need for intellectual development of the field. In fact, the connection is closer than one might think, since getting any substantial number of professional philosophers to commit to a field requires that it be sufficiently well recognized for their personal career needs to be feasible therein. A field that the wise old stick-in-the-muds of the department do not see as legitimate will not get you far; but to persuade *them* requires that reasonably acceptable philosophers generate a quite large number of publications in the field. This situation is at about the Catch 21 level, and all of us with tenure face some obligation to help build the field up. That means helping to set up symposia at the conventions as well as publications. Sometimes we might be able to join forces with others on that task—for example, it's time some symposia were run on The Changing Face of Philosophy, or on the connection of informal logic with ordinary language analysis; or we can run them ourselves, on Recent Work in Soft Logic; or we might try forays into the enemy's territory, as at meetings of The Association for Symbolic Logic where we could join forces with the fuzzy set theory crew, etc. Again, it takes staff work to do this and I'm not sure we're quite at critical mass. It may develop anyway, but more slowly, without planning and nudging.

A newsletter is a good start and may become a journal (as has *Evaluation News*); or it may spawn one; or one may move in with another journal on a cooperative basis—perhaps *Philosophy and Rhetoric*. Some articles can of course be placed in standard journals; this is especially true for those of us working in recognized paths, e.g., the history of the fallacies. But I suspect that both recognition and maturation will require some concentration of current efforts. Rather quickly, too, I think we need an historian, an archivist and probably a resource anthology or two. Respectability has its trappings; but they are not mere trappings, they are part of self-consciousness which is a prerequisite for a new discipline or field. Again, a godparent is almost essential, and colleagues will now rarely provide that kind of support.

The newsletter can be a great movement-builder, from the serious to the silly side—with competitions (Informal Analysis Problems), bumper stickers (Demand Justice for the Undistributed Middle Class!; Circular Arguments are Better than Square Ones, etc.). But it does take time; perhaps it's worth mentioning something of our experiences. *Evaluation News* has gone from a few typed pages to sixty 2-column typeset, pasted-up pages of nearly 1000 words each in its 4-year life, and it takes the new editor a great deal of time which I subsidise via USF and a hefty subsidy for typesetting which I provide via my publishing company. But at this point we have received a strong offer from a commercial publishing house, which will eliminate the need for subsidy other than editorial and secretarial time.

These are the realities of new movements, and I've seen a great many of them wilt under the pressure of the early years; I find the experience of getting philosophy teaching going as a sub-field of philosophy to be a particularly illuminating one but similar problems emerged in parapsychology, artificial intelligence, moral education, cognitive psychology, curriculum theory and several other fields in which I've been involved. Ours is a very important field and those of us that can help it in these early years will help a great many people who can benefit from its earlier flowering. For this and all the other reasons given in earlier pages, I'm proud to be present with you all at this First International Informal Logic Event.

AFTERWORD

J. Anthony Blair
& Ralph H. Johnson

The purpose of the Windsor Symposium on Informal Logic at which eight of the papers collected in this volume were presented was to serve as a stepping stone in the development of this field as a separate, if not autonomous, area of logical inquiry. In several ways this purpose has been realized.

For many of those who attended the Symposium itself, it served to stimulate an active interest in the field of informal logic.

An immediate and direct result of the Symposium was the creation of the *Informal Logic Newsletter* (which we edit and publish at Windsor), designed to be an informational organ for philosophers and others working in the field. Now in its second year, the *Newsletter* has begun publication of short articles and critical reviews, and its expansion into a journal seems imminent. It circulates to about 200 people and institutions in Canada, the United States, Britain, Western and Eastern Europe, Australia, even the People's Republic of China, and the subscription list continues to grow.

The publication of these proceedings is yet a further step. The many problems raised in these papers, and the implicit controversies their juxtaposition exposes, should provide an impetus for further philosophical activity.

This volume exhibits some of the self-consciousness of work that lies outside the philosophical mainstream. At the same time it is a step toward bringing informal logic into that mainstream.

A BIBLIOGRAPHY OF RECENT WORK IN INFORMAL LOGIC

Ralph H. Johnson & J. Anthony Blair

INTRODUCTION

1. *Organization.*
This bibliography is divided into three main sections: 1. Monographs, II. Articles, III. Textbooks. This division corresponds to the order in which the literature is reviewed in our chapter, "The Recent Development of Informal Logic," pp. 3-28. Reviews of monographs and textbooks are listed as Part B of sections I and III, respectively.

The principle of ordering in this bibliography is basically chronological rather than by the more customary principle of alphabetical ordering by author. Section II is divided into two parts: first, articles appearing between 1953 and 1968 are listed alphabetically by author; second, from 1968 on there is a year by year breakdown, and for each year the articles are listed alphabetically by author. Section III.A. is also divided into two parts: the first contains textbooks published between 1946 and 1969, the second contains those published from 1970 to 1978. In each group the texts are listed in chronological order, though where more than one was published in a given year, they are listed alphabetically by author.

2. *Principles of Inclusion.*
I. Monographs. Only those monographs were included which were judged to be focussed primarily and directly on informal logic. Thus Strawson's *Introduction to Logical Theory* (1952) and Rescher's *Dialectics* (1977), for example, were excluded.

II. *Articles.* Only those articles were included which were judged to have a direct bearing on informal logic. This is the section of the bibliography which is most problematic, since clear criteria of identity for informal logic do not exist. We were guided by the principle that a bibliography that errs on the side of over-inclusiveness can be as unhelpful as one that is too restricted in its listings.

III. Textbooks. Any textbook which devotes at least a section to informal logic (e.g., a section of informal fallacies) has been included. Textbooks restricted to formal logic were excluded. Thus we include Copi's ***Introduction to Logic*** (1953) but do not list his ***Symbolic Logic*** (1954).

3. Acknowledgments.

We would like to thank the many people at the Symposium who suggested additions to the original version of this bibliography, which was circulated there. Most particularly we are grateful to Professors Nicholas Griffin and Robert Ennis for their contributions in this regard.

I. A. Monographs 1946–1978

[1] Toulmin, Stephen. ***The Uses of Argument.*** Cambridge: Cambridge University Press, 1958.

[2] Johnstone, Henry W., Jr. ***Philosophy and Argument.*** University Park, Penn.: The Penn State University Press, 1959.

[3] Natanson, Maurice, and Johnstone, Henry W., Jr., eds. ***Philosophy, Rhetoric, and Argumentation.*** University Park, Penn.: The Penn State University Press, 1965.

[4] Anderson, Jerry M., and Dovre, Paul J., eds. ***Readings in Argumentation.*** Boston: Allyn and Bacon, Inc., 1968.

[5] Perelman, Ch., and Olbrechts-Tyteca, L. ***The New Rhetoric: A Treatise on Argumentation.*** Trans. John Wilkinson and Peircell Weaver. Notre Dame: University of Notre Dame Press, 1969.

[6] Hamblin, C. L. ***Fallacies.*** Londo n: Methuen and Co. Ltd., 1970.

I. B. Reviews of Monographs

Toulmin, ***The Uses of Argument*** (1958). [1]

[7] Brockreide, Wayne, and Ehninger, Douglas. "Toulmin on Argument: An Interpretation and Application," ***Quarterly Journal of Speech*** 46:44–53 (February 1960).

[8] Castaneda, Hector Neri. "On a Proposed Revolution in Logic," ***Philosophy of Science*** 27:279–92 (July, 1960).

[9] Cooley, J. C. "On Mr. Toulmin's Revolution in Logic," ***The Journal of Philosophy*** 56:297–319 (March 26, 1959).

[10] Cowan, J. L. "The Uses of Argument—An Apology for Logic," ***Mind*** 73:27–45 (January 1964).

[11] Manicas, Peter T. "On Toulmin's Contribution to Logic and Argumentation," ***Journal of the American Forensic Association*** 3:83–94 (September 1966).

Perelman, ***The New Rhetoric*** (1969). [5]

[12] Parker, Douglas H. "Rhetoric, Ethics and Manipulation," ***Philosophy and Rhetoric*** 5:69–87 (Spring 1972).

[13] Rotenstreich, Nathan. "Argumentation and Philosophical Clarification," *Philosophy and Rhetoric* 5:12-23 (Winter 1972).
Hamblin, *Fallacies* (1970). [6]
[14] Engelbretsen, George. *Dialogue* 12:151-54 (March 1973).
[15] Rossie, Jean-Gerard, *Archiv fur Geschichte der Philosophie* 55:91-5 (1973).
[16] Wilson, Barrie. *The Modern Schoolman* 51:182-84 (January 1974).

II. Journal Articles
1953-1967
[17] "Logic and Marketplace Argumentation," Ray Lynn Anderson and C. David Mortensen, *Quarterly Journal of Speech* 53:143-51 (April 1967).
[18] "More on the Fallacy of Composition," Yehoshua Bar-Hillel, *Mind* 73:125-26 (January 1964).
[19] "A Note on Informal Fallacies," Richard Cole, *Mind* 74:432-33 (January 1965).
[20] "An Appraisal of the Watson-Glaser Critical Thinking Appraisal," Robert H. Ennis, *Journal of Educational Research* 52:155-58 (December 1958).
[21] "A Concept of *Critical Thinking,*" Robert H. Ennis, *Harvard Educational Review* 32:81-111 (Winter 1962).
[22] "Operational Definitions," Robert H. Ennis, *American Educational Research Journal* 1:181-201 (May 1964).
[23] "Begging the Question," Oliver A. Johnson, *Dialogue* 6:135-50 (September 1967).
[24] "Some Reconsiderations of the *Argumentum Ad Hominem,*" Dale Reipe, *Dasharna International* 6:44-47 (July 1966).
[25] "The Fallacy of Composition," W. L. Rowe, *Mind* 71:87-92 (January 1962).

1968
[26] "Begging the Question," M. E. Williams, *Dialogue* 6:567-70 (March 1968).

1969
[27] "Communication, Argumentation, and Presupposition in Philosophy," Peter A. Schouls, *Philosophy and Rhetoric* 2:183-99 (Fall 1969).

1970
[28] "The Burden of Proof," Robert Brown, *American Philosophical Quarterly* 7:74-82 (January 1970).
[29] "Philosophy and the *Argumentum Ad Hominem,*" Henry W. Johnstone, Jr., *Revue Internationale de Philosophie* 24:107-11.
[30] "Philosophy of Education: Learning Theory and Teaching Machines," Michael Scriven, *The Journal of Philosophy* 62:896-908 (November 5, 1970).

1971
- [31] "Assertion Logic and the Theory of Argumentation," Leo Apostel, *Philosophy and Rhetoric* 4:92–100 (Spring 1971).
- [32] "On Begging the Question at Any Time," Robert Hoffman, *Analysis* 32:51 (December 1971).
- [33] "On Making Sense in Philosophy and Rhetoric (A Reply to Professor Schouls)," John King-Farlow, *Philosophy and Rhetoric* 4:42–47 (Winter 1971).
- [34] "Professor Schouls' Presuppositions," Gary M. Kodish, *Philosophy and Rhetoric* 4:48–54 (Winter 1971).
- [35] "Begging the Question," Richard Robinson, *Analysis* 31:113–17 (March 1971).
- [36] "Arguing from Ignorance," Richard Robinson, *The Philosophical Quarterly* 21:99–108 (April 1971).

1972
- [37] "Arguers as Lovers," Wayne Brockreide, *Philosophy and Rhetoric* 5:1–11 (Winter 1972).
- [38] "A Theory of Contextual Implication," Richard Corliss, *Philosophy and Rhetoric* 5:215–30 (Fall 1972).
- [39] "Begging the Question," David Sanford, *Analysis* 31:197–99 (June 1972).
- [40] "Reason, Semantics and Argumentation in Philosophy," Peter A. Schouls, *Philosophy and Rhetoric* 4:124–31 (Spring 1972).
- [41] "On Fallacies," John Woods and Douglas Walton, *The Journal of Critical Analysis* 4:103–12 (October 1972).

1973
- [42] "The Fallacy of Many Questions: 'How to Stop Beating Your Wife,' " Frank Fair, *Southwestern Journal of Philosophy* 4:89–92 (Spring 1973).
- [43] "Assumption Seeking as Hypothetic Inference," Donald S. Lee, *Philosophy and Rhetoric* 6:131–52 (Summer 1973).

1974
- [44] "The Problem of Fallacies," F. Cizek, *Teoria A Metoda* 6:105–15.
- [45] "The Concept of Critical Thinking," Edward d'Angelo, *Revolutionary World* 9:49–55.
- [46] "The Believability of People," Robert H. Ennis, *The Educational Forum* 39:347–54 (March 1974)
- [47] "The Concept of *Ad Hominem* Argument in Galileo and Locke," Maurice Finocchiaro, *The Philosophical Forum* 5:394–404 (Spring 1974).
- [48] "On *Argumentum Ad Hominem,*" D. Gerber, *The Personalist* 55:23–28 (Winter 1974).
- [49] "Successful Argument and Rational Belief," Gary Iseminger, *Philosophy and Rhetoric* 7:47–57 (Winter 1974).
- [50] "Logic, Language, and Formalization," Moshe Koy, *Logique et*

Analyse 67–68:389–444 (September-December 1974).
[51] "The Logic of Analogy," William Sacksteder, *Philosophy and Rhetoric* 7:234–52 (Fall 1974).
[52] *"Argumentum Ad Verecundiam,"* John Woods and Douglas Walton, *Philosophy and Rhetoric* 7:135–53 (Summer 1974).
[53] "Informal Logic and Critical Thinking," John Woods and Douglas Walton, *Education* 95:84–86 (Fall 1974).
[54] "Authority," Gary Young, *Canadian Journal of Philosophy* 3:563–83 (June 1974).

1975
[55] "The Fallacies of Composition and Division," James E. Broyles, *Philosophy and Rhetoric* 8:108–13 (Spring 1975).
[56] "Symbolic Logic and the Appraisal of Arguments," William B. Griffith, *Teaching Philosophy* 1:3–20 (Summer 1975).
[57] "The Nature of Controversial Statements," Arthur N. Kruger, *Philosophy and Rhetoric* 8:137–58 (Summer 1975).
[58] "Are There Any Good Arguments That Bad Arguments Are Bad?," Gerald Massey, *Philosophy in Context* 4:61–77.
[59] "The Appeal to Force," Dwight Van De Vate, Jr., *Philosophy and Rhetoric* 8:43–60 (Winter 1975).
[60] "Reasoning and Threatening: A Reply to Yoos," Dwight Van De Vate, Jr., *Philosophy and Rhetoric* 8:177–79 (Summer 1975).
[61] *"Petitio Principii,"* John Woods and Douglas Walton, *Synthese* 31:107–27 (June 1975).
[62] "Is the Syllogism a *Petitio Principii?*," John Woods and Douglas Walton, *The Mill News Letter* 10:13–15 (Summer 1975).
[63] "A Critique of Van De Vate's 'The Appeal to Force,'" George Yoos, *Philosophy and Rhetoric* 8:172–76 (Summer 1975).

1976
[64] "The Fallacy of Begging the Question," John A. Barker, *Dialogue* 15:241–55 (April 1976).
[65] "Counterexamples and Where They Lead," Peter Facione, *Philosophy and Phenomenological Research* 36:523–30 (June 1976).
[66] "Recent Publications in Logic," Susan Haack, *Philosophy* 51:62–79 (January 1976).
[67] "The Justification of Deduction," Susan Haack, *Mind* 85:112–19 (January 1976).
[68] "Truth, Belief and Vagueness," Kenton F. Machina, *The Journal of Philosophical Logic* 5:47–78 (February 1976).
[69] "Inadmissible Arguments," B. Peppinghaus, *Logique et Analyse* 29:119–39 (March 1976).
[70] "Fallaciousness without Invalidity," John Woods and Douglas Walton, *Philosophy and Rhetoric* 9:52–4 (Winter 1976).
[71] *"Ad Baculum,"* John Woods and Douglas Walton, *Grazer Philosophische Studien* 2:133–40.

1977

[72] "*Argumentum Ad Hominem:* from chaos to formal dialectic," F. M. Barth and J. L. Martens, *Logique et Analyse* 20:76–96 (March-June 1977).

[73] "Rescuing 'Begging the Question,'" J. I. Biro, *Metaphilosophy* 8:257–71 (October 1977).

[74] "Contextual Analysis: An Approach to the Study of Philosophic Arguments," Maurice Cohen, *Metaphilosophy* 8:3–20 (January 1977).

[75] "Reply to Woods and Walton's '*Ad Hominem, Contra* Gerber,'" D. Gerber, *The Personalist* 58:145–46 (April 1977).

[76] "Logics," William Sacksteder, *Logique et Analyse* 20:42–66 (March-June 1977).

[77] "The Fallacy of Begging the Question: A Reply to Barker," David Sanford, *Dialogue* 16:485–98 (September 1977).

[78] "A Pre/Post Test for Introductory Logic Courses," Donald Scherer and Peter A. Facione, *Metaphilosophy* 8:342–47 (October 1977).

[79] "*Ad Hominem*," John Woods and Douglas Walton, *Philosophical Forum* 8:1–20.

[80] "*Ad Hominem, Contra* Gerber," John Woods and Douglas Walton, *The Personalist* 58:141–44 (April 1977).

[81] "*Post Hoc, Ergo Propter Hoc,*" John Woods and Douglas Walton, *The Review of Metaphysics,* 30:569–93 (June 1977).

[82] "*Petitio* and Relevant Many-Premissed Arguments," John Woods and Douglas Walton, *Logique et Analyse* 20:97–110 (March-June 1977).

[83] "Towards a Theory of Argument," John Woods and Douglas Walton, *Metaphilosophy* 8:298–315 (October 1977).

III. A. Textbooks

1946 to 1969

[84] Black, Max. *Critical Thinking.* New York: Prentice-Hall, Inc., 1946. 402 pp.

[85] Searles, Herbert L. *Logic and Scientific Methods: An Introductory Course.* New York: The Ronald Press Co., 1948. (2nd ed., 1956. 378 pp.)

[86] Werkmeister, W. H. *An Introduction to Critical Thinking: A Beginner's Text in Logic.* Lincoln, Nebraska: Johnson Publishing Company, 1948. 663 pp.

[87] Hepp, Maylon H. *Thinking Things Through: An Introduction to Logic.* New York: Charles Scribner's Sons, 1949. 455 pp.

[88] Beardsley, Monroe C. *Practical Logic.* Englewood Cliffs, N.J.: Prentice-Hall, Inc., 1950. 580 pp.

[89] Copi, Irving M. *Introduction to Logic.* New York: The Macmillan Company, 1953. (5th ed., 1978. 590 pp.)

[90] Ruby, Lionel. *The Art of Making Sense: A Guide to Logical Thinking.*

Philadelphia and New York: J. B. Lippincott Company, 1954. (3rd ed., revised by Robert E. Yarber, 1974. 185 pp.)
[91] Little, Winston W.; Wilson, W. Harold; and Moore, W. Edgar. *Applied Logic.* Boston: Houghton Mifflin Company, 1955. 351 pp.
[92] Fearnside, W. Ward, and Holther, William B. *Fallacy: the Counterfeit of Argument.* Englewood Cliffs, N.J.: Prentice-Hall, Inc., 1959. 218 pp.
[93] Schipper, Edith Watson, and Schuh, Edward. *A First Course in Modern Logic.* New York: Henry Holt and Company, Inc., 1959. 398 pp.
[94] Emmet, E. R. *Handbook of Logic.* New York: Philosophical Library Inc., 1960. 236 pp.
[95] Salmon, Wesley C. *Logic.* Englewood Cliffs, N.J.: Prentice-Hall, Inc., 1963. (2nd ed., 1973. 150 pp.)
[96] Angell, Richard B. *Reasoning and Logic.* New York: Appleton-Century-Crofts, Meredith Publishing Company, 1964. 625 pp.
[97] Carney, James D., and Scheer, Richard K. *Fundamentals of Logic.* New York: Macmillan Publishing Co., 1964. (2nd ed., 1974. 428 pp.)
[98] Rescher, Nicholas. *Introduction to Logic.* New York: St. Martin's Press, 1964. 360 pp.
[99] Barker, Stephen F. *The Elements of Logic.* New York: McGraw-Hill Book Company, 1965. (2nd ed., 1974. 337 pp.)
[100] Gordon, Donald R. *Language, Logic, and the Mass Media.* Toronto: Holt, Rinehart, and Winston of Canada, Ltd., 1966. 120 pp.
[101] Freeman, David Hugh. *Logic: The Art of Reasoning.* New York: David McKay Company, Inc., 1967. 330 pp.
[102] Moore, W. Edgar. *Creative and Critical Thinking.* Boston: Houghton Mifflin Company, 1967. 340 pp.
[103] Terrell, D. B. *Logic: A Modern Introduction to Deductive Reasoning.* New York: Holt, Rinehart and Winston, Inc., 1967. 355 pp.
[104] Kilgore, William J. *An Introductory Logic.* New York: Holt, Rinehart and Winston, Inc., 1968. 352 pp.
[105] Ennis, Robert H. *Ordinary Logic.* Englewood Cliffs, N.J.: Prentice-Hall, Inc., 1969. 151 pp.
[106] Ennis, Robert H. *Logic in Teaching.* Englewood Cliffs, N.J.: Prentice-Hall, Inc., 1969. 520 pp.
[107] Kahane, Howard. *Logic and Philosophy: A Modern Introduction.* Belmont, California: Wadsworth Publishing Company, Inc., 1969. 449 pp.
[108] Michalos, Alex C. *Principles of Logic.* Englewood Cliffs, N.J.: Prentice-Hall, Inc., 1969. 433 pp.

1970 to 1978

[109] Michalos, Alex C. *Improving Your Reasoning.* Englewood Cliffs, N.J.: Prentice-Hall, Inc., 1970. 127 pp.
[110] Capaldi, Nicholas. *The Art of Deception.* Buffalo: Prometheus Books, 1973. (Formerly published by Donald Brown Inc., New York, 1971.) 192 pp.

[111] Kahane, Howard. *Logic and Contemporary Rhetoric: The Use of Reason in Everyday Life.* Belmont, California: Wadsworth Publishing Company, Inc., 1971. (2nd ed., 1976. 252 pp.)
[112] Purtill, Richard L. *Logical Thinking.* New York: Harper & Row, Publishers, 1972. 157 pp.
[113] Brody, Baruch A. *Logic: Theoretical and Applied.* Englewood Cliffs, N.J.: Prentice-Hall, Inc., 1973. 280 pp.
[114] Byerly, Henry C. *A Primer of Logic.* New York: Harper & Row, 1973. 560 pp.
[115] Thomas, Stephen N. *Practical Reasoning in Natural Language.* Englewood Cliffs, N.J.: Prentice-Hall, Inc., 1973. 280 pp.
[116] Annis, David B. *Techniques of Critical Reasoning.* Columbus, Ohio: Charles E. Merrill Publishing Company, 1974. 157 pp.
[117] Ehninger, Douglas. *Inference, Belief, and Argument, An Introduction to Responsible Persuasion.* Glenview, Illinois: Scott, Foresman and Company, 1974. 186 pp.
[118] Kaminsky, Jack, and Kaminsky, Alice. *Logic: A Philosophical Introduction.* Reading, Massachusetts: Addison-Wesley Publishing Company, 1974. 334 pp.
[119] Pospesel, Howard. *Propositional Logic: Introduction to Logic.* Englewood Cliffs, N.J.: Prentice-Hall, Inc., 1974. 211 pp.
[120] Baum, Robert. *Logic.* New York: Holt, Rinehart and Winston, Inc., 1975. 516 pp.
[121] Flew, Antony. *Thinking Straight.* Buffalo, N.Y.: Prometheus Books, 1977; published by Fontana, Great Britain, 1975, as *Thinking about Thinking.* 127 pp.
[122] Barry, Vincent E. *Practical Logic.* New York: Holt, Rinehart and Winston, 1976. 384 pp.
[123] Blumberg, Albert E. *Logic: A First Course.* New York: Alfred A. Knopf, 1976. 462 pp.
[124] Ehlers, Henry J. *Logic: Modern and Traditional.* Columbus, Ohio: Charles E. Merrill Publishing Company, 1976. 245 pp.
[125] Engel, S. Morris. *With Good Reason, An Introduction to Informal Fallacies.* New York: St. Martin's Press, 1976. 154 pp.
[126] Geach, Peter. *Reason and Argument.* Oxford: Basil Blackwell, 1976.
[127] Manicas, Peter T., and Kruger, Arthur N. *Logic: The Essentials.* New York: McGraw-Hill Book Company, 1976. 498 pp.
[128] Munson, Ronald. *The Way of Words: An Informal Logic.* Boston: Houghton Mifflin Company, 1976. 437 pp.
[129] Scriven, Michael. *Reasoning.* New York: McGraw-Hill Book Company, 1976. 250 pp. (Formerly published by Edgepress, Berkeley, 1976.)
[130] Simco, Nancy D., and James, Gene G. *Elementary Logic.* Encino and Belmont, California: Dickenson Publishing Company, Inc., 1976. 305 pp.
[131] Carter, K. Codell. *A Contemporary Introduction to Logic with Appli-

cations. Beverly Hills: Glencoe Press, 1977. 179 pp.
[132] Johnson, Ralph H., and Blair, J. Anthony. *Logical Self-Defense.* Toronto: McGraw-Hill Ryerson, 1977. 236 pp.
[133] Fogelin, Robert J. *Understanding Arguments: An Introduction to Informal Logic.* New York: Harcourt Brace Jovanovich, Inc. 1978. 351 pp.
[134] Girle, Roderic A.; Halpin, Terence A.; Miller, Corinne L.; and Williams, Geoffrey H. *Inductive and Practical Reasoning.* East Brisbane: Rotecoge, 1978. 226 pp.
[135] Runkle, Gerald. *Good Thinking: An Introduction to Logic.* New York: Holt, Rinehart and Winston, 1978. 352 pp.
[136] Weddle, Perry. *Argument: A Guide to Critical Thinking.* New York: McGraw-Hill Book Company, 1978. 192 pp.

III. B. Reviews of Textbooks

The Art of Deception, Capaldi (1971). [110]
[137] Fox, Michael, *Dialogue* 14:167–68 (March 1975).
[138] Moore, J. T., *Teaching Philosophy* 1:72–4 (Summer 1975).
[139] Rein, Irving, *Philosophy and Rhetoric* 8:194–95 (Summer 1975).
Logic and Contemporary Rhetoric, Kahane (1971; 2nded., 1976). [111]
[140] Cogan, Robert, *The Journal of Thought* 9:200–01 (July 1974).
[141] Cogan, Robert, *Teaching Philosophy* 1:477–81 (Fall 1976).
Introduction to Logic, Wilson (1971).[1] Prepared by Susan Wilson. Bletchley, Bucks., G.B.: Open University Press, 1971. Vol. I (Parts 1-9), 152 pages; Vol. II (Parts 10-18), 88 pages.
[142] Hausman, Alan, *Teaching Philosophy* 1:194–200 (Fall 1975).
A Primer of Logic, Byerly (1973). [114]
[143] Walton, Douglas, *Teaching Philosophy* 1:62–4 (Summer 1975).
Fundamentals of Logic, Carney and Scheer (1973; 2nd ed.). [97]
[144] Goosens, William K., *Teaching Philosophy* 1:68–9 (Summer 1975).
Practical Logic, Barry (1976). [122]
[145] Walton, Douglas, *Teaching Philosophy* 2:81–84.
With Good Reason, Engel (1976). [125]
[146] Walton, Douglas, *Teaching Philosophy* 2:81–84.
The Way of Words, Munson (1976). [128]
[147] Walton, Douglas, *Teaching Philosophy* 2:81–84.
Reasoning, Scriven (1976). [129]
[148] Finocchiaro, Maurice, *The Review of Metaphysics,* 30:773–75 (June 1977).
Thinking Straight, Flew (1977). [121]
[149] Walton, Douglas, *Dialogue* 17:582–3.
Logical Self-Defense, Johnson and Blair (1977). [132]
[150] Kielkopf, Charles, *The New Scholasticism* (forthcoming).
[151] Michalos, Alex, *Dialogue,* 17:584–5.

Footnotes

[1] Wilson's text really falls outside the perimeter of our survey, since it does not include a section on informal logic. However, Hausman's review of the text makes some interesting points and observations of interest to the informal logician. That is why we have included it in our bibliography.

COLOPHON

Designed and published by Michael Scriven.
Design assistance and paste-up by Bob Cooney.
Typeset by Sienna S'Zell and Bonnie Wolohan
Set at Edgepress on Mergenthaler Linoterms in Palatino 10/12,
   titles and cover in Raleigh and Cartier.
Printed and bound at GRT Book Printing, Oakland, California,
   on 60 lb. Bookwhite.